The Otter and his Hawk

A Warrior's Approach to Surviving WW2

Robert H. Sabet and Danny J. Hoskins

The Otter and his Hawk
Copyright © 2021 by Robert H. Sabet

All rights reserved. No part of this book may be reproduced or transmitted in any form or by any means without written permission from the author.

ISBN 978-1-304-07399-0

Dedication

I would like to thank the members of Grey Otter's family, for their assistance and for the information they provided regarding him, for the book. I would like to thank Danny J. Hoskins for teaching me the hawk system discussed in this book. Thank you to the members of the Rough And Tumble Society for their support. Thanks to all the Veterans past, present and future for serving our country and protecting our freedoms.

Table of Contents

Foreword 6

Introduction 12

The Iroquois Spiked Tomahawk 15

Peter Lagana 19

The VTAC 24

History of the Tomahawk 27

Tactical Tomahawks 31

Grey Otter 35

Confederate Indians 43

Moccasin Rangers? 49

Some Controversy 54

Harley Swiftdeer Reagan 62

Swiftdeer/Otter Hawk System 71

James Smith 98

One Skunk 109

Miles S. Horn 111

Ernest E. McLish 116

Bushmasters 123

Brummett Echohawk 130

Kenneth Scissons 139

Indian Scouts 149

Grey Otter's Knife Act.. 158

Grey Otter, Later Years 162

What It Means To Me 170

Foreword

In the days of my youth, I spent summers rambling the woods and ridges of Wrightwood, California, in the mountains above Los Angeles, where my folks had a cabin. Those wonderful summers were highlighted in the little village by "Mountaineer Days" celebrations over the Fourth of July and a season-closer over Labor Day weekend. For those celebrations, a Western and Native American art gallery featured performances of knife and tomahawk throwing by George "Skeeter" Vaughan, better known in this role as Grey Otter.

Obsessed then, as now, with frontier lore, I never missed a performance, dazzled by the flashing blades that Grey Otter wielded with casual precision.

Vaughan was a Western showman and a Hollywood actor and stuntman — and he was also a warrior, one of many Native Americans who served with skill and distinction in World War II. That's a tale that unfolds in these pages.

The through-line of my lifelong study of frontier history is "continuity and persistence." The qualities that made the great Frontier Partisans of yore remain relevant, indeed, vital — even in our "post-modern" world.

The tomahawk is a practical symbol of that continuity and persistence. A tool and weapon that evolved in the North American borderlands 300 years ago remained a viable and useful tool and weapon in the 20th century and on into the 21st. If you need evidence, note that the exquisite 'hawks created by Daniel Winkler are prized elements of the loadout of many Tier One operators in the Global War on Terror.

"Otter and His Hawk" explores the martial applications of the tomahawk that persisted well beyond the frontier era — but most importantly it tells stories that should never be forgotten, honoring brave men who brought their heritage and the capabilities it graced them with to bear to serve this nation when the chips were down.

I like to think that, in some Frontier Partisan Valhalla, Grey Otter is well pleased.

Jim Cornelius
Author of "Warriors of the Wildlands: True Tales of the Frontier Partisans," and keeper of the virtual campfire at www.frontierpartisans.com.

In 2008 Congress declared the Friday after Thanksgiving to be Native American Heritage Day. This is because the people who would be honored on the day *"have volunteered to serve in the United States Armed Forces and have served with valor in all of the Nation's military actions from the Revolutionary War through the present day, and in most of those actions, more Native Americans per capita served in the Armed Forces than any other group of Americans."*

These people, my Ancestors, have served with distinction in every major conflict for over 200 years with a per capita involvement higher than that of any population to serve in the U.S. military. But what can you expect from a people whose way of life, day in and day out demanded the warrior, the soldier, and the militant?

Since before the days of Columbus and Hernan Cortez even set foot on the soil of what is today the Americas, warfare has been a part of an ancient tradition. A tradition that was sometimes cruel, decisive and bloody, but necessary, nonetheless. The interpretation of the warrior varied among tribes as someone who was a protector of the elderly, children, the sick and the community, and sometimes as guardians of the ancient ways. Among others, it meant, to do what was necessary for the survival and prosperity of the people. Religion demanded it, the gods sometimes demanded it, and sometimes revenge was cause for it. 'Killing for the sake of killing' which is a concept not unique to any single group.

For thousands of years, warfare took place and it was trial and error, adaptation and the introduction and discovery of new materials that led to the creation of weapons and fighting concepts better suited to the circumstances of the times. Hand knapped knives evolved into steel blades, battle axes into tomahawks, atlatl darts to arrows, and arrows to bullets...The Native Americans by the time of European arrival were well trained and practiced in their own arts, comparable in effectiveness to the martial methods found all over the globe.

When the land finally fell to the hands of the American government, those who survived years of genocidal attempts of eradication, joined forces with the American government to protect and defend the land from outside threats, as is the sacred purpose of the Native American warrior. From World War I, World War II, Korea, and Vietnam, into the present day wars and conflicts, they have served and continue to serve with distinction. Today they are recognized for their tremendous service to this nation.

The martial concepts which existed in the old days further changed following the arrival of the Europeans, this time for the fight on the frontlines... With these warriors a martial tradition followed that proved effective, time and time again, against even the most modern weapons of warfare, impressing dictators, generals and the most ruthless enemies of the age. I leave such intimate details of history and use to be discovered further in the pages of this book. With that being said, a lineage of information has been passed down from the hands of these warriors, down to the hands of my Instructor Danny Hoskins, and now down to me as I continue to learn, write and research, as I go along.

Of Lakota and Scottish descent and a relation of Black Bird (Zintkala Sapa) of the Standing Rock Sioux Tribe, I began my journey in Arizona among the Navajo people in which I shared a home, upbringing and ceremonial participation with. With close ties to the culture I moved to the Midwest where my grandparents lived, and was homeschooled onward.

As an adult I always had an interest in my background and heritage but never knew how to explore that side of myself, even as my Lakota grandfather had been forced into the loss of most of his culture.

Eventually the death of a family member caused me to look into martial arts as a distraction to grief, which in turn led me to Master Danny Hoskins school where he taught a wide range of martial arts, including Native American Warrior Arts (NAWA) which included the dances, games, history, and fighting methods of the Native Americans.

After trying it, I have dedicated myself to it for around seven years now. I also furthered my research by writing two books on the history of Native American warfare, and practice to this day with hopes of passing down this information to my future children should I have them and whoever destiny intends to cross my path. I hope to honor my Grandfather Russell B. Dillon Sr (1932 - 2020) by revitalizing a lost tradition into our family.

As for Danny and his martial art system as a whole, I was always very interested in the weapons side of things, physically it always felt practical for my body, but the arts themselves are not limited to this.

Its lineage can be traced as far back as Niño Cochise in the time of Geronimo and Nana with methodologies coming from the Cheyenne, Cherokee, Blackfeet, Sioux, Omaha, Timucua, Seminole and many other indigenous peoples.

On the battlefield and even into World War II these methods have been utilized, especially by people like Skeeter Vaughan/Grey Otter and later Swift Deer who passed down these arts through NAWA's lineage. The tomahawk dances and applications as taught by Grey Otter and Swift Deer's twisted hairs approach were especially fascinating and effective aspects of the art. From my first-hand experience training these art forms, It cannot be doubted that such methods were a terror in real combat.

These martial arts, however, are in danger of disappearing outside of a handful of practitioners worldwide. And so It cannot be stressed enough, the importance of passing down this knowledge for future generations, as a way of honoring the Warriors who fought long before our time, their sacrifice and everyday realities, as well as the future generations who will have lost a complete and all encompassing, effective martial practice, a cultural facet and an often unexplored history.

I write this in great honor and respect to those fallen warriors, Code Talkers and service members and in recognition of the warriors who fought long before us when the Native Americans had no unifying name.

~ Bethany J. Dillon

Author of War-Torn: A Look at Warfare in North America Before European Influence

bethanyjunedillon@gmail.com

Introduction

"He cried the relief he felt at finally seeing the pattern, the way all the stories fit together—the old stories, the war stories, their stories—to become the story that was still being told. He was not crazy; he had never been crazy. He had only seen and heard the world as it always was: no boundaries, only transitions through all distances and time."

Leslie Marmon Silko - Ceremony

Hitler reportedly admired the martial tradition of American Indian tribes with warrior societies. Purportedly it was he who stated that "the most dangerous of all American soldiers is the Indian.... He is an Army within himself. He is the one American soldier Germany must fear."

I have written two books on the subject of hand-to-hand combat and the instructors who taught it, I have researched several others. I thought it would be interesting perhaps for some to know that there are still some "systems" out there which are yet to be discovered. People are not going to find anything earth shattering in these systems, but they are interesting because they provide a glimpse to the past.

I have attempted to learn these techniques whenever I am able to find someone who can teach them to me. I am not a fan of any one particular system or instructor of WW2 era hand to hand combat. Most of the time they are Judo based systems such as what arose from the New York Dojo. I wrote previously about a man named Allan C. Smith who taught Judo to the troops in both World Wars. I also wrote about a wrestler named Ed

Don George who taught the cadets of the Navy Pre-Flight School in Chapel Hill, North Carolina, his system of hand-to-hand fighting.

That's how I wound up meeting Danny Hoskins. Through my research and the writing of the two books I ended up hosting a group of like-minded individuals interested in similar things. I met Danny via the group, he is the keeper of one such unique system. Danny is a person who is dedicated to preserving the Native American Warrior Arts. It turned out that he teaches a tomahawk fighting system which was taught to him by Harley "Swiftdeer" Reagan. Reagan learned that system from Skeeter Vaughan, otherwise known as Grey Otter. Grey Otter devised the system in WW2. When I found out about this it piqued my interest and I wanted to know more about it and all that it entailed.

Danny points out that although this tomahawk system was created by someone who had Native American blood, initially he was embarrassed by it because it was nothing akin to the other Native American fighting styles he had learned and taught. It was something different, something simplistic, but also deadly and effective. He didn't feel like it was something he wanted to teach because he felt that others would question the simplicity of the system. As we have seen with other systems, during the same era, for it to be effective it had to subscribe to the K.I.S.S. principle. Keep It Simple Stupid.

According to Danny, his instructor met Grey Otter when they were working together as stuntmen. Grey Otter is primarily remembered today for his acting roles, stunt work and for being known as a master knife and tomahawk thrower.

This book initially became my attempt at understanding the man and the "system" he passed on. A system which was devised during the dark days of the Second World War.

The Iroquois Spike Tomahawk

Again, to be perfectly clear, this book is not about a "Native American" Martial Arts System per se. It is about my journey looking to find one to learn as a martial artist and instead finding something more interesting, the history and evolution behind one unique system that was devised by a man who was half Cherokee.

In 2014 I asked my college pal Rob Ludgate, who is Native American, about a man named Brummett Echohawk. It had been years since I had trained in any formal martial arts system. Several years prior I had gravitated toward Combatives and reality based self-defense systems. In particular WW2 Combatives.

Brummett Echohawk was a Pawnee who became a hand to hand combat instructor during the 2nd World War, because he excelled at the Judo which was being taught. Echohawk was also an artist who became famous for his depictions of incidents that he had lived through during the war.

I told Rob that I had always been interested in learning more about Native American martial systems and was curious if he knew anything about them. I was always skeptical of anyone claiming anything to be Native American. Martial Arts wasn't really Rob's thing or mine back when we used to hang in college, although I had been taking lessons in Judo. We used to discuss the bands we were both into. We were musicians. But I trusted Rob's opinion. Rob plays bass in an indigenous band today called Khu.eex.

In 2014 we were catching up, and it was on my mind so I figured I would ask. I told him I had been searching around for a Native American Martial Arts system to learn but most of them seemed bogus.

I told Rob about Brummet Echohawk and told him that I assumed when it was mentioned that he was an expert in hand-to-hand fighting that it was the Combat Judo which they were teaching the military back then. I told him I wanted to know where I could find more information about actual *Native American* Martial Arts.

Rob told me "Man I'd be weary of people teaching "native martial arts." I'm sure there are traditional fighting styles from tribe to tribe but I've never heard of anything ever being taught like Judo style. Honestly, a lot of it is family fighting style that that gets passed down within extended family and not super formal. There's a really long history of Native boxers going back over a hundred years. A lot of younger guys are more into MMA, but boxing has been the bigger thing historically."

Rob told me he never heard of Brummett Echohawk "but I'm familiar with the Echohawk family from Pawnee. There's a really badass artist our age named Bunky Echohawk." Rob told me "there's TONS of Native WW2 vets even though they're up there in age and passing on. Natives have the highest per capita of military service of any group in the country. If you go to any Native event there's pretty much always vets there representing."

I left it at that. I felt it was the best answer I was going to get on the subject. I focused on other things. Later that year I went to a historical event at Fort Delaware in Sullivan County, New York. One of the reenactors at one of the booths had several tomahawks on display at a table. I gravitated toward one that I thought looked really cool. The man told me it was an Iroquois spiked tomahawk which he claimed his father had found in a barn in the 1950s. When I picked that thing up I thought it was the most vicious thing I ever saw.

Later I bought a throwing tomahawk at and bait and tackle shop. I had fun throwing that thing around. I never did quite become an expert with it but I did okay. Eventually I got bored with the whole thing. I never felt like it was something I was going to ever use other than for entertainment.

Peter Lagana

One year I bought an American Tomahawk Company Lagana VTAC or Vietnam Tactical Tomahawk. Peter S. Lagana was another WW2 Veteran who designed a tactical tomahawk which was used during the Vietnam conflict. Again I thought it was the most vicious thing I ever saw. I basically put it away and brought it out every now and again to hold but that was the extent of it. I never attempted to throw it. The thing was too damn sharp.

I knew Lagana had written two manuals which I eventually got ahold of. One was a tomahawk and knife throwing manual, the other was basically a manual on using the tomahawk for fieldcraft. There was also some very interesting film footage of Lagana utilizing the tomahawk in staged hand to hand encounters. It piqued my interest again. One day while reading an old internet page I saw a mention that Lagana had been working on a tomahawk combatives manual. I made inquiries but nothing ever turned up. I found an internet forum with a post from 2002. A man named Andy Prisco who had founded the ATC Company which sold the updated version of Lagana's tomahawk wrote about a meeting with Lagana.

Prisco wrote: "After a few drinks, I probed Peters past as I never had before asking him relentlessly about the roots of his experience in hand-to-hand combat, which I learned was founded in WILL more than SKILL. He avoided the details for quite some time, but I broke him FINALLY!!!!"

Prisco goes on to write "Peter as it turns out, beyond his bloodline as a Mohawk Indian, is also a proud Judo and Jiu-Jitsu man! In Peter's words, Karate is just showmanship. Jiu-Jitsu can maim and kill. Judo defends against Jiu-Jitsu. We learned both!!! He also went on to say that few people will tear out an eye, rip a mouth apart at the lips, or bury a Tomahawk in someone's head but men of his time would, could, and did.

I also learned that Peter was asked to be a full-time civilian Hand to Hand instructor at both Fort Bragg, as well Quantico. He declined both offers of employment after thorough consideration, as his home was Western Pennsylvania. He said that while he would have loved the warmer weather of the South, he simply could not leave family, friends, and equally important, his brothers at the VFW.

In one of his CQB demonstrations featuring the Vietnam Tomahawk, he expressly told two Marines with un-sheathed, fixed bayonets to attack him as hard as they could, and even taunted them to do so, claiming that THEY were the ones who were going to get hurt when he did this demo at Quantico, it was in front of 18 officers. He bloodied his right hand during the attack, but completely disarmed two Marines attacking him from two different directions, who had complete permission to gouge, stab, and cut him at full speed.

Following the double Bayonet attack, the single Machete and single Knife attacks seemed like child's play. In fact, the knife attacker lost the knife in one second, as Peter SMASHED his hand instantly upon employment of the attacker's technique, causing the knife to fly across the field.

Now, mind you, I have only seen Peter in person as a fit, but aging, senior citizen, so it has been hard for me to envision exactly what he looked and moved like as a man in his 40's, training Marines and Soldiers in the use of the Tomahawk, let alone Hand to Hand Combat! Well, I got to see the 8 mm film footage later in the evening at Maria's guest house. I could not believe my eyes! It was like watching Rudolf Nureyev on steroids with a Tomahawk, Peter was an incredibly fit, wide-chested, wiry man jumping, rolling, and flipping all over the place! He moved with incredible agility and speed, we are trying to digitize this footage for our website.

In the film, he decimates two wooden man-like Targets with large, sweeping horizontal attacks with the Tomahawk. Then, two full speed scenarios are played out where Peter's M-1 is jammed and he must fight two attackers with a Tomahawk for his life.

After quickly dispatching the first attacker with a mock blow to the torso, he dispatches the second by throwing the Tomahawk right next to the attacker's head, in a well-positioned man sized target board right next to the attacker, for illustration (because if he wanted to, he could have placed it in the LIVE target). The smiles of the assistants after each scenario are impressive, as they have complete confidence in Peter's ability and control!"

The films of Lagana which Prisco mentioned in the forum were the ones I had seen on the ATC Company website.

TESTED AND PROVEN IN VIETNAM

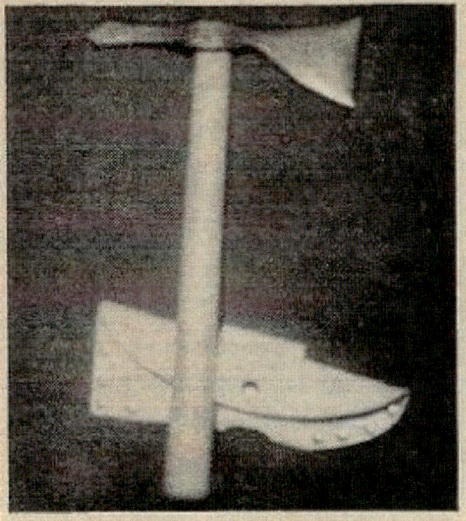

* TOMAHAWK

Proven to be extremely more effective than any knife or bayonet.

"There is no doubt that the TOMAHAWK is a formidable weapon for individual combat. Particularly when made to this design."

- Weight—One Pound
- Head Drop Forged
- Hickory Handle
- Carrying Case

$9.75 plus $1.00 Handling Charge

American Tomahawk Co.
Ebensburg, Penna. 15931

* Patent Pending

The VTAC

An article about Lagana written for the Pittsburgh Press, on February 25th 1968, provides some more background on Lagana and his tomahawk:

"The Ebensburg "chopper" may not win the war in plenty of good licks, its' inventor reports.

This chopper travels only short distances through the air and is hand-propelled. It is, in short, a tomahawk. Fittingly enough, this hair-raising weapon was designed by a great-grandson of a full-blooded Iroquois brave.

He is Peter LaGana Jr., 41, a mailman, who has been teaching use of "silent weapons" – knives, swords, hatpins, etc.and hand-to-hand combat for 23 years. He is also a gunsmith.

Naturally, Mr. LaGana is one mailman dogs instinctively shy away from. But to get back to the chopper, Mr. LaGana designed a one-pound version so delicately balanced that nobody could miss with it, he reports.

To prove this, he reports, he lets men, women and children of all ages and sizes heave it at targets 15 to 20 feet away. In 870 of these random throws, the weapon stuck in the target each time, he says.

When word of this whiz-bang missile reached the Marine Corps, Mr. LaGana was invited to demonstrate it at the Landing Force Development Center at Quantico, Va.

According to published accounts, he demonstrated its versatility against a rifle butt, a machete, a Bowie knife and a clubbed rifle—as well as a bayonet charge by two men.

The Marine officers were impressed, Mr. LaGana recalls, and each of the 18 watching bought one of the weapons. But the corps itself declined to adopt it as standard equipment. A Marine Headquarters spokesman in Washington said a major reason for this was that use of a tomahawk requires its user to lay down his rifle— and the corps doesn't like to encourage that.

But individual Marines and their relatives and friends have bought 90 per cent of the 700 tomahawks Mr. LaGana has sold over the past seven months, he reports.

And glowing reports of their multi-purpose usefulness have been drifting back to Mr. LaGana by mail—and even in personal visits. One man reported killing four Cong in hand-to-hand combat with it after his rifle had been snatched away—another chops pole-sized trees with it for quick clearing of helicopter landing fields—another chops his way through walls of huts with booby-trapped doors."

History of the Tomahawk

In the book Arms and Armor in Colonial America 1526-1783 by Harold L. Peterson, he writes that in 1675, the state of Connecticut issued an order stating that:

"Ten good and serviceable hatchets be provided in each county for the use of the army, and ten soldiers to carry them instead of swords."

Peterson writes that this was "the first known indication of a trend that was to become much more popular in the next century. The hatchet was originally adopted for use in clearing away obstructions, and the ten soldiers for each county would thus act as pioneers for the rest of the force. As the American colonists became more used to the fighting techniques of the hatchet, however, many grew to prefer them to swords." The author goes on to mention that when the individual had to provide his own weapons, it was much more practical to have a hatchet since it could also be used for domestic purposes. The hatchet became an increasingly popular substitute for the sword.

Peterson later points out that in Massachusetts, in 1779 the following order was directed to the militia:

"You are hereby ordered and directed to complete yourself with Arms and Accoutrements, by the 12th instant, upon failure thereof you are liable to a fine of Three Pounds: and for every sixty days after, a Fine of Six Pounds.

Articles of Equipment

A good fire arm, with a steel or iron ramrod, and a spring to retain the same, a worm, priming iron and brush, and a bayonet fitted to your Gun, a scabbard and belt therefore, **and a cutting sword or tomahawk or hatchet**, a Pouch containing a cartridge box, that will hold fifteen Rounds of cartridges at least, a hundred buckshot, a Jack Knife and Tow for wadding, six flints, one pound of powder, forty leaden Balls fitted to your Gun, a Knapsack and Blanket a Canteen or wooden Bottle sufficient to hold one Quart."

In The Ax Book by D. Cook the author discusses the history of the use of tomahawks in America. Cook writes: "The first type of metal ax used in North America was undoubtedly the "trade" ax, a poll-less ax with a narrow but flaring bit. Most trade axes were probably of French manufacture, though they were used in the New World trade with the Indians by both Spanish and English certainly, as well as the French. These axes were already scattered among the Indians before there were any English settlements in North America at all. Since an iron ax was such a revolutionary tool to the Indians, their desire for the metal ax guaranteed its spread, for the Indian tribes were at war with one another much of the time. Captain John Smith found trade axes among the Indians near Jamestown, Virginia, in 1607. Over two generations previously, Hernando de Soto, the Spanish explorer, had found trade axes among the Indians of what is now southeastern United States.

The probable origin of these earliest noted trade axes was the even earlier arrival of white traders to the New World in the years following Columbus.

Some of the first were Portuguese, but they were followed by the French. Many returned for fish but soon entered the fur trade also. That the trade ax was one of the more popular items of exchange was demonstrated by its spread among Indians who were far from any contact with the whites. Later English traders in the Hudson Bay region even left that ax pattern, modified somewhat, the name Hudson Bay. It was the trade ax pattern that still later was also made in the smaller and popular tomahawk size.

By the time the English colonists were established, they soon had reason to notice the distinctive axes the Indians carried. Under primitive conditions, the Indian ax or tomahawk was a small tool with its stone head lashed to a forked, or split and lashed, stick. Living without metals until the coming of the Europeans, particularly the irreplaceable iron for tools, the Indians seldom attempted to cut or use the heavy construction timbers favored by the white men. But forest life required a cutting tool. The lighter, more portable tomahawk served for the less demanding cutting and hacking the Indians needed it for. Like our own ancestors, the Indians also used the tomahawk for hacking people as well as tent poles. This way of living was part of Indian existence. To save their own skulls, they had to smash others.

The tomahawk was one of the most efficient weapons the Indians had. In reliability, effectiveness, and ease of use, the tomahawk was superior to the bow and arrow, even if not as long-ranged. *Being most vulnerable, the skull was the foremost tomahawk target.* With the sometime encouragement of white allies, it is not unlikely that this fact could have helped the grisly practice of scalping become a victory symbol. Even in our serene society, exultant spectators have been known to swarm to the field after a victorious football game and literally tear opponent's goalposts into splintered souvenirs. In some manner and degree, we are all influenced by whomever we encounter. During the settling of North America, the harsh life of the frontier whites did not differ greatly from the Indians' way of life. Forest living required a cutting tool, and to those on the move, the light Indian ax made sense. Soon white trappers and fighting men carried tomahawks of the new, reduced-size trade patter, and even ones edged with steel. So to the accompaniment of their guns, the implacable white men used tomahawks just as the Indians did. The primitive Indians were powerless to stem the never-ebbing tide of whites. By this century, few Indians were left, scattered across the continent in desolate reservations. Their tomahawk pattern has survived. You can still find a modern version, usually with a longer handle, in some northern stores. It is variously known as a "cruising," "trapper's," or "Hudson Bay" ax.

Tactical Tomahawks

In The Brooklyn Daily Eagle, Sunday, December 30th 1900, C. W. Mason in an article titled Wars To Be Waged As Of Old, predicts: "That a hundred years hence, we will have reverted to savage methods of warfare, using tomahawks and shotguns." He writes "I prophesy that in the year 2000 we shall have reverted almost to the methods of savage warfare... The most powerful nation will be that which musters the largest number of "toughs"— I mean men who are tough in nerve as well as body; that is to say, men as near the beasts as may be.

The strategy of campaigns will be the strategy of Genghis Khan, of Attila—
The overrunning of a country by millions of undisciplined robbers; the tactics of battle, hand to hand slaughter at close quarters by means of night attacks or tremendous sacrifice. In other words, I believe the education and civilization we are coddling with such pains is simply preparing us to become the victims of some new race of Goths and Huns and Tartars, from China or Africa, or the slums of our own huge cities.

Such a forecast will strike even those most ignorant of military art as a paradoxical absurdity. They must, however, remember that the Romans were scientific soldiers when the empire succumbed to Barbarians, and that large British armies have suffered defeat from small bodies of Boers organized virtually on Zulu models.

Strategy will always consist primarily in concentrating superior numbers on an isolated enemy. This depends wholly on mobility, which depends on commissariat and transport: food and shelter to keep the soldier well, and carriage to save his strength for actual fighting by relieving his body of weight and his legs of labor."

Mason later discusses the type of warfare which he predicts will take place and writes:

"They are a mile from the nearest outposts of the enemy. In noiseless boots, carrying their tomahawks in their hands, unencumbered with overcoats or haversacks or anything that jingles, they advance in close columns until their skirmishers hear the sentries.

Then very stealthily they extend right and left, in companies of one hundred, by lanes and paths, always feeling for the enemy with scouts. Signals being disallowed, a fixed time is agreed on for the real advance; perhaps as much as two hours, during which the 10,000 men would be extended over a front of four or five miles. At the hour agreed on, each company sends forward ten scouts, extends forty men in twos, and holds fifty in reserve.

The scouts find the outposts and signal by small lanterns behind them. An attempt is then made to surprise the sentries without noise, by stabbing them or throwing tomahawks."

A few years later than C.W. Mason predicted, in 2003, David Tillett wrote an article for ABC news titled Some U.S. Troops Choose Historic Tomahawk. In the article he wrote that members of the Air Force security groups, Army Rangers and Special Forces are some of the U.S. troops who have chosen to add tomahawks to their basic gear.

Tillet then asks the question, why would a member of today's armed services want a relic of the American frontier? He writes that according to one modern tomahawk manufacturer, the reasons soldiers carried them in the Revolutionary War are still valid today —it all comes down to science.

Ryan Johnson, the owner of RMJ Forge states that "The physics behind it make it an appropriate choice for any kind of battlefield conditions."

Johnson continues "You take a knife, a knife has a certain amount of leverage that's given to you. The tomahawk can be used like a knife, but you also have that 18 inches of handle that gives you a huge amount of difference in power as far as the power of the cutting stroke. It's much more practical as a field tool because you can again use it like a knife or you can use it like an ax."

Johnson then goes on to state that soldiers have used tomahawks in most of the major wars the United States has fought. He says "In World War II, there were not only Native Americans using them, but also you're your regular GI. A lot of these people were just carrying stuff from home, stuff that they used on the farm."

Johnson added that he had an uncle who served in the Korean War. He told him that soldiers would take the standard hatchet that they were issued and grind the back down into a spike to make a "fighting hatchet."

As was the case with Lagana and his tomahawk during the Vietnam War, the article by Tillet goes on to mention that not everyone was sold on the tomahawk's potential for widespread acceptance in the military. He writes That retired Army Maj. Gen. William Nash, a military analyst, said the Army is not quick to add new items, and add weight to the list of gear that a soldier has to carry. Safety is also a concern and commanders often have reservations about providing soldiers with untested items, or allow them to carry one that they purchased themselves.

Nash says "I've been in outfits where any private weapon—to include knives—were not permitted. But as the lethality of the weapon increases, the tolerance for its presence decreases. They become too unaccounted for."

In comparing the tactical tomahawk to the entrenching tool, Nash says "It's hard enough (to dig a fighting position) with an entrenching tool. The hatchet's a better hatchet than the entrenching tool is. But we didn't buy the entrenching tool for a hatchet. We bought it to dig holes."

He goes on to say "Now... at the same time, an innovative person comes up with something that may be useful, but it takes a long time for the Army to test it and get it in the field. That frustrates the soldier."

Johnson from RMJ Forge states that in his opinion the tomahawk won't be a standard-issue item for all of the military, but "I think it will definitely be an issue item for a lot of the Special Forces eventually."

Grey Otter

Grey Otter, otherwise known as George E. "Skeeter" Vaughan, was taught how to throw both the tomahawk and the knife by his Cherokee grandfather Limping Bear. His grandfather used them when he was a warrior for the Confederacy during the Civil War when he fought under the great Indian General Stand Watie.

I was able to contact Grey Otter's family members and they were gracious enough to offer input toward this section of the book. I also quoted heavily from Harry K. McEvoy's book Knife & Tomahawk Throwing, however where things he wrote were in dispute I introduced the new information. It's important to set the record straight.

According to McEvoy, he states that Grey Otter's grandfather began teaching him when he was eight years old. The education "dealt mostly with a large range of weapons including rifles, revolvers, and pistols, with special emphasis placed on knives and tomahawks. Considerable time was also devoted to the handling of whips for driving mules and horses, along with the use of ropes for basic ranch work."

Grey Otters' daughter, who will remain anonymous out of respect for her privacy, was able to confirm that it was Grey Otter's grandfather who taught him most of his technique.

McEvoy mentions that Grey Otter mastered all those things with skill and expertise but he developed a special affinity for knives and tomahawks and the art of throwing them. During the Great Depression, Skeeter used them to provide wild game for the family table at a time when he couldn't afford rifle and shotgun ammunition.

During his formative years, he worked on ranches in various states in the West, as a cowboy and horsebreaker. He also did several stints with rodeos as a performer. One year he worked in "Doc Willetson's Big Medicine Show" as it toured the West. He threw his tomahawks and 16-inch-long knives made from old British bayonets.

At the age of 19, in 1942, he enlisted in the U.S. Cavalry, taking basic training at Fort Riley, Kansas. "He was soon elevated to the position of a full-time instructor. Besides teaching recruits the use of weapons, he also trained them in unarmed combat—a field in which he was considered an expert, despite his being of just a medium build and standing only five feet eight inches in his combat boots." Grey Otter's daughter noted that "Skeeter was actually closer to 5'6 and stocky, although he was leaner during the war."

During his two years at Fort Riley he reached the rank of sergeant, "during which time he taught literally thousands of recruits the basics of weaponry, from handguns, rifles, and carbines, to machine guns and the 37mm cannon."

McEvoy writes that Grey Otter was later transferred to the 18th Recon. Squadron located at Fort Lewis, Washington. He soon ended up in England, in the spring of 1944. He landed with his unit on Omaha Beach in Normandy, two days after D-Day, as part of the Allied invasion forces. However, his daughter disputes this and stated that he was part of an advanced recon unit and was there before the Normandy invasion. To her knowledge he was never stationed at Fort Lewis, so this information seems to be in dispute as well.

"It was about two weeks later that the unit of Indians called the Moccasin Rangers was organized. This was a very elite group of 15 men, all of whom were trained and blessed with "night sight." It was organized for night patrols and reconnaissance work against the Germans and usually operated behind enemy lines. The exploits of this special unit soon became well-known, for it was a force that the Nazis found difficult to deal with."

Back to the tomahawk system. When I learned that Danny was teaching this system I made inquiries. I had heard of Grey Otter before because of a crazy story in which he threw a knife an extremely far distance in order to remove a sentry. McEvoy describes the incident as follows:

"The guns were silent as sounds of intermittent shelling faded away in the distance. There was no moonlight, but the night skies were clear and full of stars, and what light there was reflected off the mantle of newly fallen snow. Visibility was quite good, considering the darkness of this late November night in 1944, and the blackish object in the distance could be seen clearly against the snow.

That dark object was a German sentry, on guard duty beside a bunker complex which included a Nazi pillbox that had held up the American advance all day and was giving the U.S. Infantry considerable resistance.

The six-man patrol of the U.S. Moccasin Rangers—all American Indians especially gifted with "night sight"—was commanded by Sergeant George E. Vaughan, nicknamed "Skeeter." At the moment, the squad was belly down in the snow behind the German bunker complex, concealed some 35 yards away in a small stand of timber on the brow of the hill.

The sentry had his back to the hill and stood facing the American lines. He was at least ninety feet downhill from where the Indian G.I.'s lay in the snow assessing the situation.

The six-man patrol had silently filtered through the German lines and circled behind the bunker-pillbox fortress. Their mission was to take out the pillbox by eliminating the garrison, and to accomplish the job before dawn when the Allied advance was to resume. A great many American lives depended upon the success of this mission.

The problem at the moment was to eliminate the sentry and to do it silently. It would be easy to shoot the soldier, but the crack of a rifle shot would alert the Germans and make this important assignment a costly, if not an impossible, one. It would also be very difficult to creep down from the timber line and knife the sentry, since it was all open ground between them. Furthermore, since the German seemed quite alert, an assailant might be quickly spotted approaching against the light background of new snow."

McEvoy writes "There was a slight chance that the problem could be solved by the commander of the patrol, a full-blooded Cherokee Indian from California named Skeeter Vaughan. He was only one month away from his 22nd birthday, an event he might never celebrate if his plan failed to succeed." However, this information is incorrect according to his daughter, she informed me that Grey Otter was not a full-blooded Indian but was of part Cherokee and Welsh ancestry.

McEvoy continues "Skeeter Vaughan and his patrol lay in the snow in the midst of the German army's fabled Siegfried Line, and the bunker and pillbox below were a part of the vast network of barriers to the Allied advance.

"What are we gonna do, Sarge?" asked a G.I.

Skeeter shook his head silently, his eyes intent on the back of the sentry 30 yards away.

"Could you throw your knife that far and nail him? We're gonna have to try something," the G.I. insisted.

"O.K.," said Sgt. Vaughan. "But even for a pro, that's one hell of a long throw—and all downhill too! If I miss, you guys better be ready to blast him, just in case."

He drew the weapon from its sheath and held it by the blade. It was a customized knife he had made from a bayonet—16 inches in overall length and exactly the type of throwing-knife he had used most of his life, on and off the stage.

Quietly Skeeter crawled out of the timber as far as he dared and stood up. The back of the sentry was still toward him. Now Skeeter was in the open, silently praying that he had been unobserved.

With a skill developed over many years of continual practice, Skeeter hurled his bayonet knife in a high trajectory, aiming for a spot about three feet above the head of the sentry. The weapon turned silently over and over in its long downhill pinwheel flight, and to Skeeter's amazement, the sentry dropped face down into the snow without a sound—the weapon had penetrated the sentry's head at the base of his skull. Had Skeeter missed, or only slightly wounded the sentry, his Moccasin Ranger team would have had to open up with rifles or automatic fire, thus alerting the pillbox. The entire mission might then have been a failure.

As a result of Skeeter's miraculous throw, however, the patrol was able to reach the back door of the pillbox unobserved. They knocked gently on the door and the German soldiers inside, thinking their relief had arrived, opened up—and were quickly eliminated by the Moccasin Rangers."

It's important to note here that according to his daughter Grey Otter was "the only person not amazed that he made the shot."

McEvoy continues, "The next day, the Americans advanced further into Germany, and members of Skeeter's unit, checking tracks in the snow, measured Skeeter's life-and-death knife throw at 87 feet—downhill and in the dark. Considering the distance thrown, Skeeter had probably made the longest successful throw in the annals of knife throwing, thereby adding another exploit to the legend of this fine Cherokee Indian who is now recognized by his peers as the world's greatest professional tomahawk thrower and the equal of any knife thrower in the business."

Danny told me that everything he was given about the Grey Otter system was from his instructor Swiftdeer, when he was in his early 20's and Danny admits that he did not keep any notes back then.

"To be honest I did not know till many years later that Skeeter was such a big deal. Some of the things I remember grandmaster Reagan saying was that he preferred to wear moccasins over his combat boots when he went out on covert missions. That he was a very good boxer." Danny also added "He once struck someone across the face with his hawk and pretty much cut his face off."

According to Grey Otter's daughter Grey Otter was not a boxer but rather an unarmed combat specialist as McEvoy had stated. She pointed out that "he never harmed anyone outside of combat." She also mentioned and McEvoy mentions this as well "Skeeter got trenchfoot and nearly had his feet amputated. He put on two large boots and hiked back to his unit. His feet were too sensitive for boots after the war, so he wore mocs from then on. But he only wore standard issue boots while in the Army. (He was 18th Cav)."

On Friday December 12th, 1980, an article in The San Bernardino County, reported that Grey Otter was going to be the grand marshal in Yucaipa's annual Christmas parade, it also notes that he taught knife and tomahawk throwing to the military. The article states "Grey Otter, also known as Skeeter Vaughan, served as the leader of the "Moccasin Rangers" of World War II and trained instructors in knife and tomahawk throwing for the Korean and Vietnam wars."

Confederate Indians

In The Life Of General Stand Watie And Contemporary Cherokee History written by Mabel Washbourne Anderson in 1931 there is a section titled Tribute To General Watie's Indian Brigade. That section provides some background on what Grey Otter's grandfather was involved in during the Civil War.

"General Watie's First Cherokee Regiment was composed of Cherokees and white men; principally of the former, but there were many brave men from Arkansas and Missouri, who enlisted under him, and whose devotion to him was equal to that of his kindred race. This famous First Cherokee Regiment was the nucleus from which was formed at the latter part of the war, "Waties' Celebrated Indian Brigade." General Stand Watie and his Indian Brigade marched over more miles, took part in more battles, had more independent conflicts and skirmishes, captured more trains of wagons and more horses and mules than perhaps any one brigade west of the Mississippi. With his forces he resisted almost step by step the advance of the Federals into the Cherokee Nation. The bodies of these men marked the way like human milestones. Many of them sleep in unknown and forgotten graves, but the memory of their valiant deeds in defense of the South is a heritage that every patriotic Southerner should be proud to cherish.

All these Indians, though untrained in the tactics of war, were by nature and habit well-fitted to become ideal soldiers. They were good riders and splendid marksmen, most of them having been athletes from their youth up, trained in self-control and endurance. Tireless in their march upon the enemy, they played well their part in the great struggle of the South from 1861 to 1865. Their acts of valor and endurance would fill no small volume. It is a historical fact that the Five Civilized Tribes lost more men in action in the Confederate Army—in proportion to the number enlisted—than any other Southern State. These Tribes furnished fully six thousand men to the Southern troops. At the close of the war a large per cent of them had given up their lives in battle. Of Watie's own brigade there are less than one hundred men living today. While there were no really great battles fought on Indian Territory soil, yet this country certainly has a war history; for the eternal vigilance required to protect the Cherokee border from the invasion of the enemy, and the skirmishes that took place almost continually were really worse than open battles.

Most of the Oklahoma text-books, on which the public school children depend for their knowledge of Oklahoma History, give little or no insight into the lives and characters of the leading men of the Five Civilized Tribes, who were the real makers of a great part of that history. However, it should be borne in mind that the unbiased historian is handicapped in his work as to details in Oklahoma History by the fact that unfortunately the Cherokees and other tribes have left no written record of their own. This is as strange as it is unfortunate in view of the fact that they produced minds gifted in statecraft, literature, oratory, and military leadership. **But like their legends and traditions many of the most valuable details of their history have been handed down from one generation to another, and the chronicler who is unable to gain this knowledge from reliable testimony must depend too much upon remote sources, such as Ethnological and Congressional Reports, which are bulky but not always dependable for real facts.**

Less has been said and written, even upon generalities, touching the Indian's part in the cause of the Confederacy than upon generalities, touching the Indian's part in the cause of the Confederacy than upon any other phase of the War Between the States. These textbooks make no mention of the Indian's part in the battle of Pea Ridge, Newtonia, Wilson Creek, and other engagements in Arkansas and Missouri in which they did such valorous service. Little has indeed been written, and that which has sometimes been quoted as hearsay concerning the Confederate warfare in the Cherokee Nation originated from those who knew least about conditions in this country at that time. Some have even dared to insinuate that among the Cherokees it was chiefly a guerilla warfare. Nothing could be more unjust or further from the truth. General Watie, before his promotion, was ever under the direction of his superior officers and was always in perfect harmony with them. Maxey, Gano, and Cooper were especially warm friends of his. Cooper had the utmost confidence in him. He said of Watie after the war "He was not only a soldier, brave, efficient and courageous, but he was a great man, whose honor and integrity were above reproach."

The fact that the Five Civilized Tribes were not required to engage in battle outside their own territory, according to the Treaty with the Confederate States, and that they did not take an active part in a number of battles in Arkansas and Missouri, is but another proof of their loyalty and devotion to the Southern cause. At the time of the War Between the States, the Indian Territory was remote and separated, and her troops so few compared to the thousands engaged in this great conflict, that this may in a measure account for the scant mention made of their services. It is not the Indian's nature to advertise their part in a victory, and no one seems to have thought sufficiently of them as factors to expend the time and means to go over the ground and gather up the interesting and valuable additions to the pages of Oklahoma's History that might have been gotten from living lips. With the stoic resolution characteristic of their race they fought and endured the perils and hardships of war in silence, and many of their deeds of valor will forever remain in oblivion."

On a website thomaslegion.net which is dedicated to Thomas's Legion which was formed during the Civil War by William Holland Thomas, who was the only white man ever to become chief of the Eastern Band of Cherokee Indians on a page titled Cherokee Indians: Weapons and Warfare, states "Cherokee weapons were designed, created, and engaged for "close-range combat," so the Cherokees, consequently, were masters of guerrilla warfare and had perfected it generations prior to the Europeans' arrival in the Americas. Hunting game had required both experience and skill. The Cherokee, through hundreds of years of practice, had adapted and transitioned its hunting prowess from "the game to the enemy."

During war, Cherokee Indians typically bivouacked at the "foot of a steep ridge," and this allowed them to: conduct hit-and-run tactics, limited the enemy in its attack formation, and it allowed them an expeditious escape route through familiar terrain.

Though the Cherokee were by nature a peaceful people, they were nevertheless trained and prepared for protecting themselves from surrounding tribes and, later, from the white man.

They became expert weapon-makers. Arrows crafted from flint and eagle feathers were secured to cane shafts and shot by bows made of sycamore and hickory. These bows were carefully shaped with bear oil and seasoned by fire. Buffalo hide breast-plates, shields, helmets and quivers adorned the Cherokee warriors while they wielded their stone tomahawks and flint-tipped spears.

Cherokee Indians developed the throwing hatchet style of the Tomahawk. (That method of fighting was lost after the Trail of Tears.) Basically, Cherokee could hunt with a special balanced hatchet. Up to a range of 30 feet, a Cherokee wielding a Tomahawk could split a coconut. In a melee, Cherokee wielding the hatchet were able to "open up the chests of those they attacked with a single blow."

Moccasin Rangers?

CLOSE FIGHTING IS THEIR DISH
Sergt. Billy (right) taking a bayonet away from private
Knight as they practice the advance arts of Judo.

I attempted to find out more information regarding who the other Moccasin Rangers were. In the book McEvoy states that as the war went on "Skeeter and his Moccasin Rangers were involved in numerous exploits that made them heroes among the American troops. Unfortunately, one by one, these extremely dangerous missions killed off his Indian friends, and by the end of the war, Skeeter was the only surviving member of the original team of Moccasin Rangers."

Although I did not find anything further regarding the Moccasin Rangers I was able to find a few articles regarding Native Americans who had served in WW2. One was an article in the Boston Globe, November 29th, 1942. The article titled "Indians Are the Best Damn Soldiers in the Army," begins with Major Lee Gilstrap loudly exclaiming this to the writer, and it continues:

"Maj. Gilstrap knows Indians. He fought beside Indians in the last war, coached them in football at Oklahoma Military Academy during the peace years, and is "Big Chief" to 2000 of them right now.

Some of the officers at this post assert that Secretary of War Stimson himself would vote the same way. They recall that Stimson was driving through the camp when his eye was caught by the feline grace and agility of an instructor in bayonet practice.

"Stop the car," ordered the Secretary. He watched in silence while the swift-moving bayonet flashed in the sun. "I want to meet the instructor," he said.

Stimson then complimented Sergt. Chauncey Matlock as "the finest instructor in bayonet practice I have ever seen"—an accolade to a full-blooded Indian who was a star football player and English scholar at Oklahoma College."

In the article Major Gilstrap says "The Indians love to use that bayonet and that probably explains why they are the best bayonet fighters." The article continues "Maj. Gilsrap's favorite example of over-use of the bayonet is that of an Indian named Hopocantubbe who served under Gilstrap in the last war. Hopocantubbe was out scouting in no man's land when he flushed a big Prussian in a shell hole instead of drilling him with a bullet, Hopocantubbe chased the Prussian for 500 yards right down the middle of no man's land and into a dugout. No shots were fired even then, but only one came out, and it wasn't the Prussian.

Out of more than 2000 Indians at this post, the records show that the only ones who have not risen above the rank of private are a few "28-day-soldiers." A 28-day-soldier is one who is good for 28 days and bad for the three days after pay day."

Gilstrap states "The Indians make such fine soldiers that they soon become non-commissioned or regular officers. We have Indian officers in all branches and they rank all the way up to lieutenant colonel."

In McEvoy's book it's interesting to note that although he was wounded five times, "and appropriately well decorated, Skeeter turned down a battlefield commission, preferring his role as sergeant. After the war, the army kept him from being discharged for several months because of his specialized knowledge and teaching qualifications. It seems the army wanted him to teach a new group of servicemen combat and command techniques before allowing him to don his buckskins and head back to California."

The Boston Globe article states that the: "most famous Indian fighter of the war so far is Maj. Gen. Clarence L. Tinker, commander of the Hawaiian Army aviation forces, who was killed in the battle of Midway.

Indians may prefer to use the bayonet, but it is a fact also that they are the best rifle shots in their division. About half of them have an expert's rating, and most of them are particularly adept at long-range rifle shooting.

"At scouting and patrol work," Gilstrap adds, "the Indian stands out like a sore thumb. During recent combat maneuvers, one Indian single-handed captured a tank and its crew; another Indian came back with 87 'scalps,' or identifying arm bands."

The sense perception of many Indians is so acute that they can spot a snake by sound or smell before they can see it. They have an uncanny faculty at weaseling over any kind of terrain at night and there is a saying that "the only Indian who can't find his way back to his own lines is a dead Indian."

This comment of course brings to mind the Moccasin Rangers and their night sight.

The Boston Globe article also mentions that the "real secret which makes, the Indian such an outstanding soldier, in Gilstrap's view is his "enthusiasm for fighting." Sergt. Echohawk, for example, a 126-pound Pawnee, is a judo expert who, in a rough and tumble battle, could snap the back of an opponent twice his size. Echohawk daily practices taking knives and clubs away from "enemies."

The article continues "This fighting spirit is attested by many semi apocryphal tales. One concerns a portly Indian who tried to join the Army and, told by the recruiting officer he was too fat to qualify, tartly replied, 'Don't want to run. Want to fight.'"

The great classic on the Indian's fighting attitude, however, was made 25 years ago by John Rat, a Cherokee. When he came home from France in the last war he was asked by his friends how he liked the Army. His answer is still echoing in this war, "too much salute, not enough shoot."

Some Controversy

The history of Native Americans in the United States and in the military is a sensitive subject which has been fraught with controversy. I attempted to navigate through it and have a better understanding of what it all entailed.

In Winona LaDuke's book the Militarization Of Indian Country she writes that during World War II: "Native service men and women described feeling "accepted as an individual, not as one of a minority group." Studies indicate that "this higher level of acceptance, particularly for reservation based Native men and women, meant that racism was not as apparent as in the border towns of home. Interviews with World War II veterans, found that some of the psychological barriers which had always been present in relations between the Indian and the White were torn down.

While explicit racism was muted in the military, it remained a constant force at home. The impact of returning to America and experiencing no change in institutional racism cannot be overstated. Many – as was the case of Navajo Code Talkers in World War II – returned and found that they were unable to vote (eleven states specifically barred Native people from voting until the late 1940s or 1950s), and had difficulty securing loans from the GI Bill. In terms of "white tape" and bureaucracy, neither the Veteran's Administration nor the banks would make a loan on allotment lands or Indian trust lands, barring any entry into the American dream.

One of the most prominent cases of this irony was Ira Hayes, a Pima man who served in World War II and participated in what quickly came to be one of the most iconic images of the war and a universal symbol of pride in military service: the flag raising at Iwo Jima.

The army pulled Ira and two other surviving men who raised the flag at Mt. Suribachi out of combat and pressed them into a bond drive campaign to raise money to pay for the war, touring the country, endlessly raising symbolic flags to replicate the event for the public.

Ira Hayes struggled with his position and the reality of America. (As a note, not a single member of the Pima people served in World War I; nineteen Pima who served in World War II died in combat.). He suffered from PTSD that the military took no interest in diagnosing or treating. He personally rejected the notion that his actions made him a hero when so many "real" heroes had given their lives. His story became legendary with Johnny Cash's recording of *The Ballad of Ira Hayes*."

LaDuke writes in her book "More than 44,000 American Indians, out of a total Native American population of less than 350,000 served with distinction between both the European and Pacific theaters of World War II."

She goes on to point out "During World War II, Native American men and women on the home front also showed an intense desire to serve their country and were an integral part of the war effort. More than 40,000 Indian people left their reservations to work in ordnance depots, factories and other war industries. American Indians also invested more than $50 million in war bonds, and contributed generously to the Red Cross and the Army and Navy Relief societies.

Laduke mentions that it is estimated that there were 42,000 Vietnam-era Native American veterans. Laduke also writes that "At the beginning of the twenty-first century, there are between 160,000 and 190,000 Native American military veterans, about 10 percent of all living Native Americans. This is a proportion triple to that of the non-Indian population. An estimated 22 percent of Native Americans 18 or older are veterans."

Echoing what my friend Rob had mentioned to me, LaDuke points out that "Today, Native peoples have the highest rates of enlistment of any ethnic group in the United States. In some Native communities, in a single graduating class over half of the graduates will be military-bound. Many youth, both reservation-based and urban, see no options outside the military to secure an economically stable future."

LaDuke asks "How did things change? How did we move from being the target of the US military to being the US military itself?" She says that the answer to the question has to do with the larger forces of American society – economic deprivation, domination and racism – all of which have figured into the high level of Native induction into the military.

She writes "Military indoctrination has long been considered to be an effective method for transforming Native America, as a way to "save the man and kill the Indian," and in many ways it has succeeded."

We can see that this dynamic goes way back. In an article written for Infantry Magazine, titled Building a Unit Combatives Program, Major Robert M. Squier discusses an incident from the Nation's history. He writes that in June of 1744, the College of William and Mary invited the Native Americans of the Six Nations to send 12 of their young men to their institution to receive a civilized education. The intention of the offer was to bridge the gap between the Europeans and the Native Americans in hope of assimilating the tribes into the growing colonial population.

The Chiefs of the Six Nations offered this reply:

"Sirs,
We know that you highly esteem the kind of learning taught in Colleges, and that the maintenance of our young Men, while with you, would be very expensive to you. We are convinc'd, therefore that you mean to do us good by your proposal; and we thank you heartily. But you, who are wise, must know that different nations have different conceptions of things; and you will therefore not take it amiss, if our ideas of this kind of education happen not to be the same with yours. We have had some experience of it. Several of our young people were formerly brought up at the College of the Northern Provinces; they were instructed in all your sciences; but, when they came back to us, they were bad runners, ignorant of every means of living in their woods, unable to bear either cold or hunger; knew neither how to build a cabin or take a deer; or kill an enemy, spoke our language imperfectly, were therefore neither fit for hunters, warriors, nor counsellors; they were totally good for nothing. We are,

however, not the less oblig'd by your kind offer; tho' we decline accepting it; and, to show our grateful sense of it, if the Gentleman of Virginia will send us a dozen of their sons, we will take care of their education, instruct them in all we know, and make men of them.

Major Squier discussing the incident then writes "The tribal leaders knew that training methods influence training outcomes. The life skills that their warrior culture demanded could not be instilled through academics; the young braves had to experience challenges, endure hardship, and overcome obstacles. The old chiefs knew that warriors are not built in a classroom. Today, that reality is unchanged. As we pursue lethality and readiness as a force, we must remember that these characteristics begin with an individual who internalizes the Warrior Ethos and commits to developing a skill set and a mindset that is combat ready."

As discussed earlier people such as Grey Otter had this mindset instilled in them from a young age. But the other thing that the article points out is the fact that there were many attempts over the course of this Nation's history to assimilate the tribes. I have to admit that this is not a subject that I have studied in depth. I know it's a sensitive topic and I have attempted to understand it as best I could over the years.

In the book Neither Wolf Nor Dog, the author Kent Nerburn writes about his encounters with Dan, a Lakota elder who unflinchingly speaks the truth about Indian life, past and present.

In one conversation Dan tells Nerburn: "Think about this. Do you ever hear white people saying that they are part black or part Mexican? Hell, no. But the world is full of people who say they were part Indian. Usually they'll say it was their grandmother or their great grandmother, it's never a grandfather. You wouldn't want an Indian man in your background. He might have had a tomahawk or something. You want some old blanket Indian woman who taught your family wise ways. And they're never a Potawatami or a Chiracahua or a Tlingit – usually it's a Cherokee. Something about the Cherokees is more romantic. I bet I've met a hundred white people who say they had a Cherokee grandmother. And you know what? They believe it! They want it to be true so much that they make themselves believe it.

'Mostly they leave it at that. But some of them don't. They grow their hair in braids and go to some powwows. Maybe take a class from some phony medicine man, and presto! We've got a new Indian. Pretty soon they're spouting Indian philosophy and twisting up the idea of the Indian even further.

I tell you, Nerburn, being an Indian isn't easy. For a lot of years America just wanted to destroy us. Now, all of a sudden, we're the only group people are trying to get into. Why do you think this is?"'

Nerburn says "I told him I didn't know."

The elder answers "I think it's because the white people know we had something that was real, that we lived the way the Creator meant people to live on this land. They want that. They know that the white people are messing up. If they say they are part Indian, it's like being part of what we have."

Out of deep respect to Grey Otter, the Native Americans, especially my Native American friends; as well as to my friend Danny and his teacher Swiftdeer, I wanted to tell this story as accurately as possible, warts and all.

There is some controversy surrounding Swiftdeer. I'm not going to recount all of it here, if people would like to find out more about that subject they can certainly go digging for it. What I will say is that there is controversy surrounding the man due to his claim that he was part Native American, Cherokee as a matter of fact. Similar to Grey Otter being half Cherokee and Welsh, Swiftdeer claimed to be half Cherokee and Irish.

Swiftdeer claimed that he grew up on reservations in Texas and Oklahoma and as was typical of many Native American families, he initially learned the skills to fight and defend himself from his Uncles and other family members.

Swiftdeer related one story to Steven Barnes in the November 1988 issue of Karate/Kung Fu Illustrated. He tells Steven about how tough his mother was "Once she went to get her husband out of a bar in Tucson. There were three whites making him drink at the back of the bar. He had gotten sick and fallen down. They were kicking and hassling him. She said, 'Can I get my husband and leave, please?' She was called a squaw and worse. Then they made the mistake of laying their hands on her,"

Swiftdeer continued chuckling grimly, "She did so much damage to those three big bruisers that the law came after her. Her family had to hide on the reservation."

Despite that controversy I felt that it didn't take away from the fact that it was entirely in the realm of possibility that Swiftdeer could have picked up Grey Otter's system from the man while they worked together as stuntmen. The system is not overly complicated and can be taught in a short amount of time. Swiftdeer was an accomplished martial artist so he would have had very little trouble learning it in a short time span. Essentially he would have wound up being in the right place at the right time, meeting someone who was willing to pass on the knowledge that he possessed.

I attempted to verify things as much as I could. It is possible that Swiftdeer made up the whole thing. Even if it was made up and Swiftdeer devised this system, it is a very practical system and we can view it as an homage to something from an earlier era. I believe it is a made up system, but I also believe that Grey Otter was the person who made it up. I will offer my reasoning in another chapter.

Harley Swiftdeer Reagan

There is no doubt that Swiftdeer knew his stuff. He was an accomplished martial artist and a former Marine. The problem with writing this book was confirming that the story of Grey Otter passing on his system to Swiftdeer was true. There is no way to resolve this, unfortunately the controversy surrounding Swiftdeer tends to cast a cloud of doubt surrounding any of his legitimate claims.

As mentioned previously, Danny said he was too young to take any notes regarding what Swiftdeer taught him and it was not until many years later that he realized the value of what he was taught. He was not aware of who Grey Otter was at the time.

Danny reached out to Swiftdeer after reading Steven Barnes article when he was 17 years old. That was in 1988. He said it was that article "that set me of on my Native American Warrior Arts journey. The next year I called Reagan in California and my research began."

Swiftdeer lived in California for a while before moving to Phoenix, and that is most likely where he met Grey Otter. Regarding the veracity of the claim that Swiftdeer picked up the tomahawk part of his fighting system from Grey Otter, Danny is sure it is true. Danny added "Nowhere have I seen Swift ever make money promoting Grey Otter's hawk."

Danny stressed to me "I know the guy was very controversial but he never took a dime from me for training and talked about many different teachers he had. I saw photos and videos with him and many other people including Grey Otter."

The problem with verifying the story as a researcher is one needs something or someone putting the two people together in the same room. I believe my friend Danny when he says he saw photos of Grey Otter and Swiftdeer together. One of the problems with tracing the lineage with a system like this is that things were not written down, as my friend Rob had told me and as the elder in Nerburn's book discussed. These systems were passed on by word of mouth. Mostly kept within families.

Nerburn writes that the elder asked him "Tell me, Nerburn, what is an Indian, anyway?" Nerburn responds "It's one of the people who was here originally." The elder then asks "Okay, Where did we come from?" Nerburn tells the old man the Bering Strait theory. This sets the old man off.

The old man says "You're afraid to say we started here, that the Creator put us here. You come out with that damn Bering Straits idea, just like this magazine."

Nerburn tells him that nobody knows. The response is "What do you mean, nobody knows? We know. But nobody believes us. We know in our hearts who we are. We have the stories from our ancestors. But we can't prove anything. If we say we are the first people, the ones who are from here, some damn archaeologist will jump up and tell us we came over through Alaska on a land bridge. They want to make sure that we're immigrants, too. Just that we got here earlier."

The old man continues: "If we say that our ancestors tell us we started here, some anthropologist will pull on his beard and tell us that is just a myth.

Then if we don't even try to talk about where we came from, but just say we are part of a tribe, no one will believe us without proof. We say we have the proof in our stories, but that's not good enough. We are told it must be written down. But the people who wrote down the tribes were all white people or Indians who worked for white people and they made all kind of mistakes.

And what about the Indian people whose tribes were destroyed and don't exist anymore? Are you going to say that those people aren't Indians because they aren't members of a tribe that the government recognizes?"

That same predicament is often what Martial arts teachers or Instructors of Reality Based Self Defense systems are faced with. People want proof that the system is real. Oftentimes this is resolved by showing some type of certification or claiming some type of lineage. However, anyone can come up with their own stories or create their own lineages.

In the case of something like the tomahawk combatives system that Grey Otter devised it's not as necessary. Many people learned varying systems of hand to hand combat systems taught by all sorts of Civilian and Military instructors during the Second World War, and they were not being taught these systems as a martial art. They were being taught these systems as a small part of a larger course of instruction which included other skills such as fieldcraft and marksmanship.

However, what is necessary to understand is that these techniques which were being taught were effective. It's a no brainer with the Grey Otter/Swiftdeer tomahawk system. It is effective.

I contacted Steven Barnes who wrote the articles for Black Belt magazine. He wrote about his association with Swiftdeer and about Swiftdeer's martial arts system.

In one article Barnes wrote titled The Core of Chulukua-Ryu American Indian Traditions and Techniques, Swiftdeer states "everyone is pretty much aware that reconnaissance means doing your homework. That's your prerequisite knowledge, your own resources, your own abilities. It's also knowing the enemy's strengths and weaknesses.

'What are you going to do? Tactics are present tense: Now it's happening, what can you do to change it?"

Barnes continues to write "In Swiftdeer's pragmatic martial philosophy any technique that drops an assailant on the street is a good technique. Of course all of us want to look good, or at least be in halfway decent form when we do something. But the bottom line is "Are your tactics deadly? You've got to learn to make your reconnaissance as swiftly as possible. Your strategy should be silent, so that the opponent is never aware of what you have in mind."

If we look at Grey Otter's system, the tomahawk fighting tactics are indeed deadly. In terms of what Swiftdeer was discussing, if we view Grey Otter's system in terms of being a complete system with the addition of the throwing techniques, the throwing techniques and the tomahawk killing techniques could have been and were used for sentry removal. An example of this would be the incident mentioned in McEvoy's book, when Grey Otter threw the knife to take out the Nazi sentry.

One other interesting thing to note here is, according to Danny, it's a misconception that the tomahawk was thrown overhead. Oftentimes the tomahawk was thrown underhanded. Also, if it was thrown overhead the thrower was not necessarily concerned that the tomahawk landed with the bladed side sticking into the opponent. If it hit him in the head or on the body with the blunt side that was sufficient, especially if it was able to cause an injury. In an article in newspaper The Spirit of Democracy, Tuesday, June 3rd 1879, Colonel Blank an Indian Fighter who was commissioned by the Confederacy as Colonel of a mounted Texan regiment describes a fight with a band of Apaches.

Of the encounter, he states: "The Apaches were taken completely by surprise, and although some of them got in the rocks, most of them were either shot or drowned.

When the fight was about over, all of a sudden I felt queer. I felt just like when a man is shut up in a dark room and can't see, and somebody comes in. He may not be able to see or hear the person, but something tells him that there is somebody near him. I never felt safer in my life than I did up there, but still I turned around to where the path was, and I saw the face of an Apache just coming above the rocks. I jumped up, and so did he. I did not have any time to get out a weapon, for I could see the flash of his tomahawk I went at him, and he threw his tomahawk. The dull edge hit me on the forehead, and it split my skull open." Blank then goes on to describe a frightful hand to hand encounter with the Apache, which he survived.

That story is interesting because it demonstrates how practical in nature Grey Otter's own system was. Despite Grey Otter being a master tomahawk and knife thrower, it wasn't necessary that one had to be skilled at making the blade land perfectly. It just had to do enough damage to the opponent to allow time to bridge the gap. One also presumes that if one were to throw a tomahawk or a knife that it would not be their primary weapon unless it was a desperation move and they damned well be sure that they were going to strike their intended target. As Frank Dean, another champion knife thrower, wrote in his book The Art Of Knife Throwing: "The idea that a throwing knife is valuable as a weapon for self-defense or offense, is erroneous for it is only practical as entertainment.

True, a great many knives are carried as weapons, some by those skilled in the art of throwing them, but rarely, if ever, are they so used. The knife fighter would be without a weapon if he threw his knife!"

This brings us back to Grey Otter's tomahawk system. It was used as an in-close system of combat. Danny said that his teacher Swiftdeer told him that Grey Otter hacked a man's face off with the weapon during the War. It's a gruesome detail, but it was War after all. It was Kill Or Be Killed! One other thing to note is that Swiftdeer himself was a Vietnam Veteran.

I contacted Steven Barnes in an attempt to confirm that the story of Grey Otter's system being transferred to Swiftdeer was true. I had already known the answer more or less. It's not something which can easily be confirmed.

Mr. Barnes wrote "I cannot absolutely verify that meeting took place, but Harley certainly said it had. That's as far as I can go on the issue, and know no one who can verify the story. Sorry!"

I thanked him for his time. I didn't feel let down. I was happy he responded. Danny reached out to Swiftdeer's wife, Diane Reagan, and she told him that she would look for photographs but she told him "I only heard about Skeeter Vaughan from Swift."

Shortly after writing the first draft of this manuscript Danny reached out to a former student of Swiftdeer; the man's name was Terry Lettau. In the May 1987 issue of Black Belt magazine, Barnes mentions Swiftdeer in his Secrets of the Shaolin Temple column. The main topic of the column is Jeet Kune Do which Terry Lettau, a 9th degree black belt in Shorei-Ryu and a 7th degree in Shorinji Ryu Jiu Jitsu, speaks of himself and his teacher and what they learned.

He states "My teacher, Harley Swiftdeer Reagan, and I both trained with Bruce, maybe two or three times a month. Jeet Kune Do never had a chance to get off the ground because Lee died too soon. He wasn't here long enough for people to really develop under it. A lot of people who say they knew what it was just don't because Lee wasn't here long enough. I trained with him, and I don't really know what it was about."

He continues on "Bruce Lee just seemed to be a step ahead of everyone else. He mastered the kung fu forms, boxing and wrestling and took the best of each, creating his own thing (jeet kune do). *He was one in a million—one of the few who could formulate a new system—but he produced no students from scratch.* We'll just never know how good Jeet Kune Do could have been."

When Danny reached out to Lettau and asked him if he had met Skeeter his answer was "Yes with teacher."

Grey Otter's family member informed me that Grey Otter had only taught his system to a few people in his life, and he regretted doing so because they all turned around and tried to use his name and techniques for profit or to pass it off as their own. This was unfortunate and I told her I was sorry to hear that. I told her I would convey the good and the bad in putting this story together. I also told her that I would not use all the information she was kind enough to provide me with but I would include this particular information, that Grey Otter regretted teaching his system, in order to provide some context. I informed her that what I was attempting to do was get the story jotted down for posterity.

Swiftdeer/Otter Hawk System

Again, it must be clear that despite my best intentions I was unable to verify the story that Grey Otter taught Swiftdeer his tomahawk system. Despite the lack of verification I believe that the system is real. One of the reasons why I believe it is real is that, on its own it's not something which would be a money maker, it's a very basic and easy to learn system, and it would not take a great deal of effort or time to learn it.

According to Danny, Swiftdeer said that Otter's movements when he utilized this tomahawk system were similar to that of a mongoose. The use of the weapon itself is a very bare bones, no frills approach. There is nothing flashy about it. Danny made it very clear to me that what he teaches is a small part of a bigger system of tomahawk fighting techniques drawn from various sources. But the Grey Otter system in and of itself is unique in the greater system of techniques that he teaches, because it is nothing like the other systems.

One of the main things to point out about the system is it utilized a double handed grip, starting from the guard position. Similar to how a soldier would wield a bayonet. From that position one of the follow-up techniques involves swinging the axe out from the guard in a double handed blow similar to how a lumberjack would chop wood, the target areas being the head and upper body, the initial blow is aimed at the neck. The follow up blow is swung out to the temple. It's basic, and it's deadly.

In McEvoy's book he writes about how Grey Otter received the nickname "Skeeter." McEvoy writes: "His now famous nickname was received when he was just a youngster when, although only 12, he was doing a man's work of bucking and falling timber in an Oregon lumber camp. One payday, the camp foreman held back a $10 bill that the boy had earned, taunted him, and told Skeeter to come and take it from him if he was man enough. The lad really needed that money.

Compared to the foreman, who was a big husky man weighing perhaps 225 pounds, Skeeter was a skinny kid, about five feet tall and around 110 pounds in weight. Skeeter started to walk away from the foreman, but after a few steps he stopped, pulled out his long throwing-knife from his belt, turned and threw it so that it passed within an inch or two of the foreman's ear. The knife stuck in a tree just behind him.

As the foreman reached behind him to withdraw the knife, Skeeter pulled another blade from his boot and waited. Suddenly the big man, pale-faced and shaky, took out his wallet, threw a $10 bill on the ground before him and walked away.

One of the lumberjacks who had witnessed the entire episode, proclaimed loudly to the group gathering around the boy: "For a little 'skeeter, he sure carries a big stinger!" And, of course, the nickname stuck!"

The main thing to note about an axe is that it is a top heavy weapon and it is primarily used for chopping. Again, everything begins from the guard position as one would wield a rifle and a bayonet.

From FMFM 1-1, Marine Bayonet Training, 25th March 1965 regarding the guard position it states:

"As in boxing the basic position of the bayonet fighter is the GUARD position. The bayonet fighter in this position is relaxed and alert. The initial attack movement begins from this position. Each movement consists of an attack and a recovery. The recovery is in fact a return to the guard position. In executing a movement the phases follow each other without a deliberate pause, thus making the entire movement a uniformly smooth action. The attack may be continued without returning to the guard position by repeating the same movement or utilizing another movement. In the guard position the bayonet fighter is ready to move into the attack to ward off his enemy."

From the book The Scout and his Axe, the author John Thurman writes. "I would recommend this as a sensible process for learning how to use a felling axe:

First and foremost, use an axe of a weight and length of haft that is suitable for you. Using an axe that is too long or too short, too heavy or too light, is unsatisfactory and can be dangerous. There are different sizes and different weights because people are of different sizes and strengths. Those of you who play cricket will know how impossible it is to play properly with a bat that is too large or one that is too small, and it is just the same with a felling axe."

The author goes on to list out the things that are important about using a felling axe, they are:

1. Grip the axe firmly throughout the whole operation of using it
2. Have your feet firmly planted on the ground. Movement should come from your arms and from your trunk above the waist.
3. Keep the axe under control throughout each stroke. You can practice swinging by having an old haft with a weight fixed firmly in place of the head. As in golf, the swing is terribly important.
4. Practice on a falling timber because it is much easier to cut a log that is resting on the ground than to cut a standing tree.
5. Learn to aim each blow at an exact spot on the log. When you are practicing, put a chalk mark on the log and try to hit that.
6. Learn to keep your head still and your eyes always on the point you are trying to hit. This, again, is exactly as in golf.
7. Learn from an early stage to use a felling axe with a right-handed grip and with a left-handed grip. This is essential because when you work in the woods there are many occasions when you have to work from the opposite side of the tree to the one you would normally choose. A right-handed grip is where the left hand is below the right hand, and a left-handed grip is where the right hand is below the left hand.
8. *Learn to let the axe do the work.*
9. *At the top of each swing the guiding (upper) hand should touch the axe head.*
10. *At the moment of impact with the log the two hands should touch each other.*

The last three steps are what Danny demonstrated from the guard position. Let the axe do the work, swing out and at the moment of impact the two hands should meet together. These are techniques of someone who was familiar with chopping wood, such as a lumberjack. Someone such as Grey Otter.

For comparison purposes in his book Woodmanship, Bernard Mason describes chopping in this way. He writes:

"An expert in the art of chopping is not born in a day. Full mastery comes only after one has had an ax in his hands for years. But the fundamentals are clear, and can be picked up by anyone, and once in hand, will do much to make the big chips fly.

Holding the Ax. – Grasp the ax with the left hand just above the knob at the end of the handle, and support it with the right hand about three-fourths of the way up the handle. With the ax held in this way, crosswise in front of the body, we are ready to start chopping.

The Forehand Swing – This is to cut the right side of the notch. Raise the ax up behind the right shoulder as in the picture. The hands are still in the same position as at the start. Now bring it down onto the log with a natural, easy, swinging motion, sliding the right hand down the handle as you do so, so that both hands are together at the end of the stroke. Raise it again, *sliding the right hand up as before*, and start the next swing. The right hand thus slides up and down, while the left remains stationary.

The Backhand Swing. – This is to cut the left side of the notch. Raise the ax over the right shoulder as before, but lean the body well to the left, so that the ax can be brought down in line with the left face of the notch.

Chop Gently. – Above all, take it easy. Never drive the ax or force it. The weight of the ax is sufficient to do the chopping. Force is unnecessary, but worse – it destroys your aim, and *accuracy is what counts.* Swing with a normal, natural, unforced, rhythmic swing – and watch your aim. That is what cuts wood, not brute force.

Danny points out that ripping is one of the major techniques in the Grey Otter system. He also discussed several other ways of gripping the tomahawk. One thing to note is that he does not advocate gripping the tomahawk at the bottom of the handle like other people show in their systems. Danny explains that one's hand will get really tired after swinging the top heavy weapon around in that manner and so it is better to grip the tomahawk more in the middle if one were to use it single handed.

Danny expanded on what he learned regarding Grey Otter's system and provided input toward the following section regarding Grey Otter's approach to gripping the hawk in combat. Danny explains the system as he learned it this way:

When I first began learning Grey Otter's approach to the hawk in 1994 with Grand Master Reagan I was taught the importance of gripping the Tomahawk and the concept of movement over technique. GM Reagan said a movement can be interpreted and used in many ways, whereas a technique is one dimensional and has limited application. There are 8 basic grips in his approach. Each grip is used to facilitate an attack or counterattack on your opponent. Grip strength is vital and the stick wrestling game is used to develop strength and weapon retention skills in the beginning. Let's dive into each grip!

THE APPROACH

SHORT GRIP

This grip is used when drawing your hawk from your belt. From the belt draw the hawk can be thrusted towards the opponent's throat with the top point of the hawk the hand is held flat palm down during this Tomahawk punch.

MIDDLE GRIP

The middle grip is the primary grip used in attacking your opponent. In this position one can deliver offensive and counter offensive movements known as ripping.

LONG GRIP

The long grip serves one purpose only in this approach and that is for power throwing. Power throwing is the concept of throwing your weapon as hard as you can toward your opponent not concerning yourself with sticking the hawk but striking anywhere you can to harm and or distract your opponent.

LANCE/BAYONET GRIP

This grip is the heart of Grey Otters World War 2 combat system. One learns the 4 movements of offensive and counter offense as well as how to finish the opponent with the coup de grâce. In this position one also learns how to turn and strike in any direction.

TWO HANDED GRIPS

The two-handed grip is used like gripping a long axe or baseball bat. From this position one can deliver a finishing blow towards his opponent. Also, from this position one can also delivery a lumber jack style throw, thrown overhead towards the opponent.

THE SEAT BELT

The seat belt is a term I coined when used to describe this grip. When first learning this grip, one is taught to hold the opponent around the neck, body or legs. After one has a tight grip around the opponent a choke, slam or takedown can be used. It was explained to me that this grip was taken from the Cherokee game stickball AKA the little brother to war.

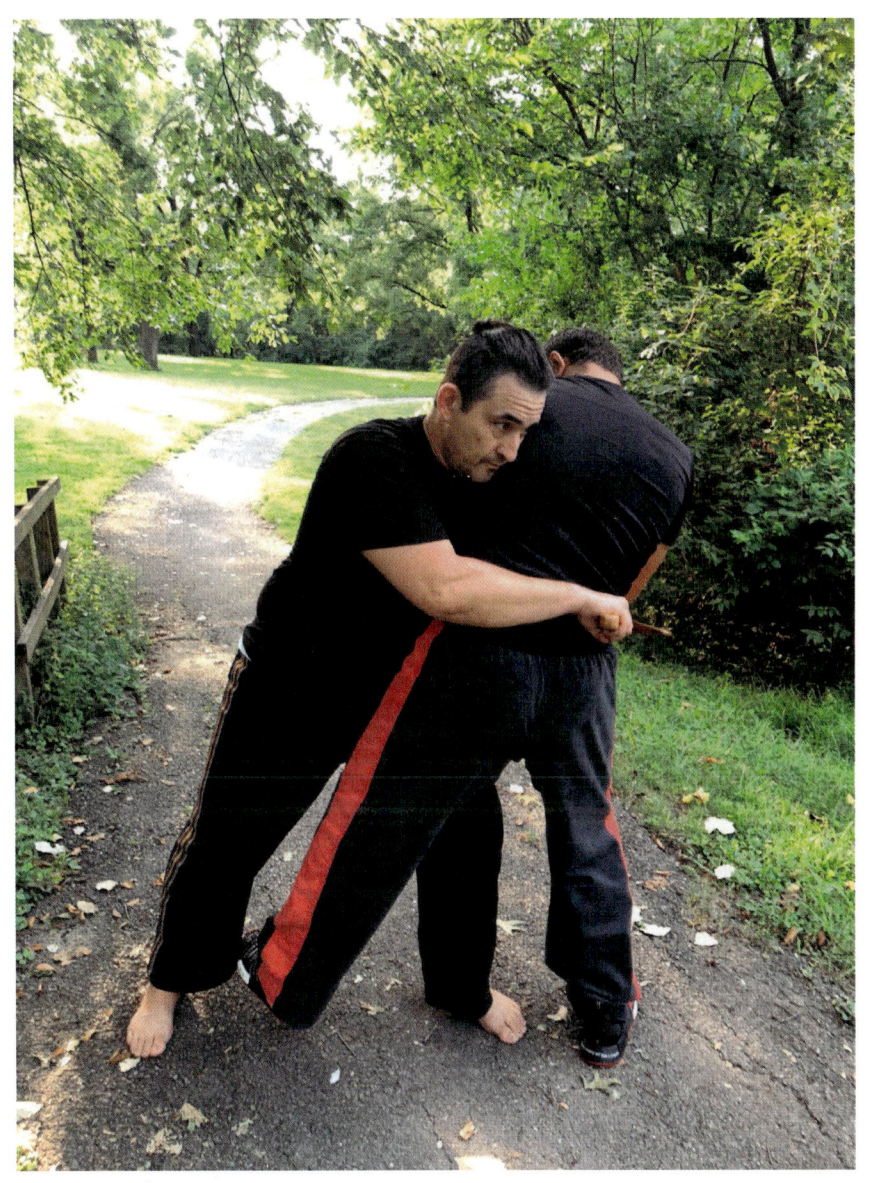

LOW GRIP

The low grip is a resting position where the hawk is held downward by your side. From here the hawk can be used to strike upward with the back portion of the head of the tomahawk towards your opponent's groin. From the low grip the blade could also be turned toward the opponent and an underhanded throw can be used to surprise an opponent.

REVERSE GRIP

The reverse grip is used in close when an opponent has closed on us and is trying to tie up the weapon bearing arm the hawk can be passed over to the opposite arm and used to strike the opponent. The reverse grip can also be used when we have slammed an opponent on the ground and are withdrawing our weapon from underneath his body to administer the finishing blow to the head in a downward hammering motion.

GREY OTTER CROSS ARM GUARD

One additional thing worth mentioning regarding the linking of Swiftdeer's system to that of Grey Otter. Danny mentioned that the ripping motion is like that of a Mountain Lion or a Bear. He wanted me to find pictures of a mountain lion using its claws in a ripping motion. At some point in time one of Grey Otter's tomahawks had been sold. I was able to locate some images of it. The back of the tomahawk offered another compelling clue that Swiftdeer had learned the system from Grey Otter.

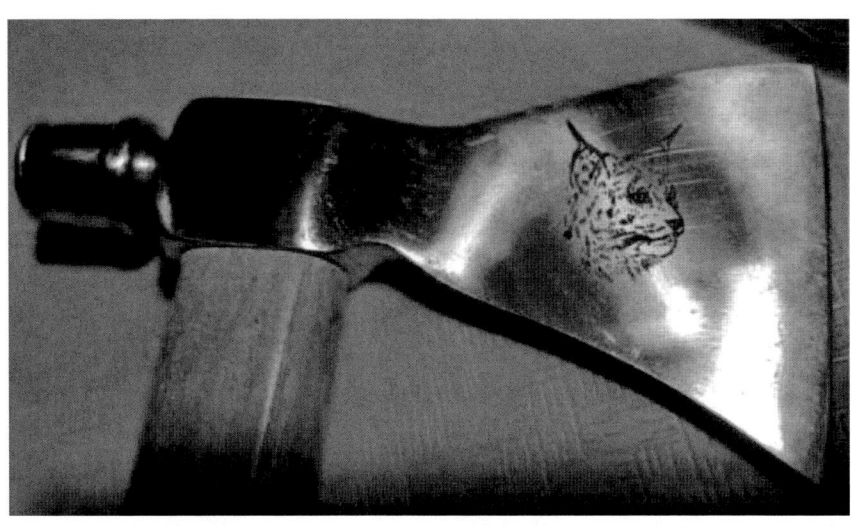

James Smith

I felt frustrated by the lack of information available about Native American fighting systems. While searching for information for this book, I became curious about their methods of warfare and how it influenced the frontiersmen who began to inhabit the United States. I did not find a lot on the subject of Native American tomahawk fighting tactics or the subject of Indian warfare but I was able to uncover some things.

One article 'A Kind of Running Fight': Indian Battlefield Tactics in the Late Eighteenth Century by Leroy V. Eid states "Indians most certainly also knew how to charge as well as retreat. Indian commanders had an important option in their use of this third additional battlefield tactic. On occasion, Indians would move forward against a large force in a grand rush. On these occasions, it was as though the gun had never been introduced. With the psychology of terror at work, this apparently rash approach was not only practical but relatively bloodless. So successful could this be that John Cleves Symmes generalized that frontiersmen must, like Indians, have a "tomehawk" so as to have the ability to charge in the woods also." Symmes was a delegate to the Continental Congress from New Jersey, and later a pioneer in the Northwest Territory.

One of the best references regarding the Native American method of warfare which I came across was written by James Smith. In 1755, Smith worked on Braddock Road, which was a road built west from Alexandria, Virginia which was in support of General Edward Braddock's ill-fated expedition against the French. Smith was captured by Delaware Indians and brought to Fort Duquesne at the Forks of the Ohio River, where he was forced to run a gauntlet before being handed over to the French.

Smith was adopted by a Mohawk family, he was ritually cleansed and made to practice tribal ways. He ultimately gained respect for Indian culture. He wrote about their methods of warfare in his memoirs.

In one section of his book titled "On Their Discipline, And Method Of War" he writes:

"I have often heard the British officers call the Indians the undisciplined savages, which is a capital mistake—as they have all the essentials of discipline. They are under good command, and punctual in obeying orders: they can act in concert, and when their officers lay a plan and give orders, they will cheerfully unite in putting all their directions into immediate execution; and by each man observing the motion or movement of his right hand companion, they can communicate the motion from right to left, and march abreast in concert, and in scattered order, though the line may be more than a mile long, and continue, if occasion requires, for a considerable distance, without disorder or confusion.

They can perform various necessary manoeuvers, either slowly, or as fast as they can run: they can form a circle, or semi-circle: the circle they make use of, in order to surround their enemy, and the semi-circle if the enemy has a river on one side of them. They can also form a large hollow square, face out and take trees: this they do, if their enemies are about surrounding them, to prevent from being shot from either side of the tree.

When they go into battle they are not loaded or encumbered with many clothes, as they commonly fight naked, save only breech-clout, leggings and mockesons. There is no such thing as corporeal punishment used, in order to bring them under such good discipline: degrading is the only chastisement, and they are so unanimous in this, that it effectually answers the purpose.

Their officers plan, order and conduct matters until they are brought into action, and then each man is to fight as though he was to gain the battle himself. General orders are commonly given in time of battle, either to advance or retreat, and is done by a shout or yell, which is well understood, and then they retreat or advance in concert.

They are generally well equipped and exceeding expert and active in the use of arms."

Later in the section Smith writes: "Though large volumes have been wrote on morality, yet it may all be summed up in saying, do as you would wish to be done by: so the Indians sum up the art of war in the following manner:

The business of the private warriors is to be under command, or punctually to obey orders – to learn to march a-breast in scattered order, so as to be in readiness to surround the enemy, or to prevent being surrounded – to be good marksmen – to learn to endure hunger or hardships with patience and fortitude – to tell the truth at all times to their officers, but more especially when sent out to spy the enemy.

Concerning Officers. They say that it would be absurd to appoint a man an officer whose skill and courage had never been tried – that all officers should be advanced only according to merit – that no one man should have the absolute command of an army – that a council of officers are to determine when, and how an attack is to be made – that it is the business of the officers to lay plans to take every advantage of the enemy – to ambush and surprise them, and to prevent being ambushed and surprised themselves – it is the duty of officers to prepare and deliver speeches to the men, in order to animate an encourage them; and on the march, to prevent the men, at any time, from getting into a huddle, because if the enemy should surround them in this position, they would be exposed to the enemy's fire.

It is likewise their business at all times to endeavor to annoy their enemy, and save their own men, and therefore ought never to bring on an attack without considerable advantage, or without what appeared to them the sure prospect of victory, and that with the loss of few men: and if at any time they should be mistaken in this, and are like to lose many men by gaining the victory, it is their duty to retreat, and wait for a better opportunity of defeating their enemy, without danger of losing so many men."

Smith sums up the section by asking: "Why have we not made greater proficiency in the Indian art of war? Is it because we are too proud to imitate them, even though it should be a means of preserving the lives of many of our citizens? No! We are not above borrowing language from them, such as homony, pone, tomahawk, etc. which is little or no use to us.

I apprehend that the reasons why we have not improved more in this respect, are as follows: no important acquisition is to be obtained but by attention and diligence; and as it is easier to learn to move and act in concert, in close order, in the open plain, than to act in concert in scattered order, in the woods; so it is easier to learn our discipline, than the Indian manoeuvres. They train up their boys to the art of war from the time they are twelve or fourteen years of age; whereas the principal chance our people had of learning, was by observing their movements when in action against us.

I have been long astonished that no one has wrote upon this important subject, as their art of war would not only be of use to us in case of another rupture with them; but were only part of our men taught this art, accompanied with our continental discipline, I think no European power, after trial, would venture to show its head in the American woods.

If what I have wrote should meet the approbation of my countrymen, perhaps I may publish more upon this subject, in a future edition."

In an article written by Russel White Bear in the Lawrence Daily Journal, dated Friday January 12, 1912, he discusses Early Day Indian Scouting. The article provides a further glimpse into the mind of an Indian Warrior. He writes the following:

"I shall endeavor to explain briefly, Indian Scouting, and their woodcraft. On learning of scouting by the Red Man you will readily see a marked distinction in it, from that of the pale face brother. This contrast lies, exclusively, in the characteristics of the Indian; for he differs from the white man, a natural born warrior, cunning in manner, a naturalist, and a skillful artist in woodcraft. The numerous ideas which the white man must acquire are natural to the Indian.

There is no code of ethics which govern Indian scouts. They are exceedingly brave, quick in thought, noted for endurance and fleet-footed. This, together with their skillful knowledge of out-door life, cunning ability of imitating birds and animals and their other many ideas to decoy their enemy, make them valuable scouts. This the United States army realized when they engaged Indian Scouts, enemies of the Sioux, to scout for the army in the conquest of the Sioux.

The scout usually gains some high point, where lying on his belly in the shadow of some tree or rock he sees everything without being seen himself; his horse meanwhile being either ricketed or grazing with dragging lariat behind the crest of the hills. The expedients adopted for concealments are many and ingenious.

The scout sometimes crawls towards a rock on the crest of a hill and when near it he draws his blanket over his head and shoulders, covering everything but his eyes and then wiggles himself by degrees up to the rock or place where he remains motionless until he has satisfactorily scanned all the country in sight, when he withdraws as stealthily as he approached, whether anything has been discovered or not.

Frequently the scout conceals himself by holding a piece of sagebrush in front of him while lying down. Again he will fasten bushes to the upper part of his clothing or body, extending above his head, then sitting in a washout or buffalo wallow he is completely concealed, while his own view is unobstructed.

Buttes and other prominent points near the trail are sought as lookouts. The scouts often go miles to reach them, unless the enemy is known to be near. In almost every case the scouts work in pairs or more. A watch is kept to the rear as well as to the front.

An Indian scout dresses in the simplest manner. Almost his entire clothing made of buckskin. With him he carries a knapsack made of hide which he holds his food. His food must be light in weight, small in bulk, that it may be easily carried, and of such substance that it will sustain life. Hence, in this sack may be found dried meat, crushed, also some fat. This meat is from some animal as buffalo, deer, or antelope. This together with water, comprise the scout's bill of fare when on an extended trip and isolated habitation.

At times it is necessary for the scout to walk; such cases he can easily cover a distance of thirty to forty miles a day.

Indian scouts are careful to never enter a new place without first reconnoitering it; and if it does not afford an easy means of escape they keep out. If a stream is reached which cannot be forded, they swim without any hesitation. Absolutely nothing escapes their notice; tracks upturned or removed stones, broken twigs or branches, ashes of camp fires, in fact they are all eyes and ears.

Now let us go back to the early day Indian and learn more about his bow and arrow, his clothing and his craft in living in the woods. First his bow is made from the ash tree. This has the two qualities most essential for a perfect bow. That is, it will bend sufficiently and has the required strength to send the arrow into the air with a tremendous force. The material for the bow is best when selected in the Spring and laid away to season, being carefully watched that it may become perfectly straight. This process requires two or three weeks' time before it is in good condition for an ideal bow.

The arrows are best when made from cherry wood, which grows extensively throughout the northwest. Like that of the bow, they are best when gathered in the Spring. Cherry saplings, as nearly as possible the desired size, are gathered, the bark removed, tied in bundles and laid away in a dry place where they undergo the same process as the bow.

When thoroughly seasoned they do not bend when striking an object; for the purpose of this seasoning is to prevent bending or breaking when hurled from the bow into the air. They are not ready to be prepared for use. Three feathers are split and proportionately attached to the head of the arrow by means of wrapping both the top and bottom of the feather with sinew. For the points, various shaped flints are used. These were later replaced by metal.

There is a kind of double sack for holding the bow and arrows used by the Indians. This is called a quiver. One division for holding the arrows is short and large while the other, for holding the bow, is much smaller but longer. Quivers are made from the skin of some animal. That of the buffalo calf being considered the best for making these.

Having referred to sinew perhaps some would like to know about this. It is the Indian's thread or cord. They sew with sinew. If a cord is needed to wrap anything, sinew serves the purpose very nicely. The bow strings are made of sinew. Sinew is taken from the flesh of some big game as the buffalo. That part of the flesh on either side of the back bone is cut out and the sinew separated from the flesh and hung up to dry. In two or three days this will be dry and will be probably, an inch and a half wide and about eighteen inches long. When needed it is torn into threats or cords as desired."

To sum it up, in Smith's other work A Treatise on the Mode and Manner of Indian War, his Indian mentor, Tecaughretango lays down the principle that "the art of war consists in ambushing and surprising our enemies, and in preventing them from ambushing and surprising us."

One final note. There is a report "Dated at Fort Edward the 25th October 1757 and humbly presented to His Excellency the Right honourable the Earl of Loudoun, Major General Commander in Chief of His Majesty's Forces in North America," which was written by Robert Rogers; he details his methods of warfare and attack with the rangers whom he is in charge of. It is found within Loudoun's papers located in the Manuscripts Department at The Huntington Library, Art Collections, and Botanical Gardens.

Rogers Rangers were a company of soldiers from the Province of New Hampshire which were raised by Major Robert Rogers and attached to the British Army during the Seven Years' War. The Rangers were quickly adopted in the British army as an independent ranger company.

Roberts trained and commanded the rapidly deployed light infantry force, which was tasked mainly with reconnaissance as well as conducting special operations against distant targets.

During World War Two, the U.S. Army was interested in the tactics of the British Commando units. Recalling the Rogers Rangers, they took the name "Rangers" as the official title. These units consider Robert Rogers their founding father.

Back to the papers. There is one section where Rogers writes: "If I was attackt in broken ground and found I could not Make a Stand to advantage

I would retreat to a better situation in an irregular scattering manner so as to make the Enemy conceive I was flying, and when I had recovered the Spot I imagined would answer our purpose, I would let them approach very close & pour in a Volley upon them & immediately afterwards those nearest in front should fall upon them with their Hatchetts, and with the rest if possible surround them in their surprize."

One Skunk

An article in the Kearney Daily Hub, Tuesday, January 26th 1943 is titled Private Oneskunk, Man of Few Words. Shows British Stalking Instructors a Thing or Two About Stalking.

The article discusses Private Sampson P. Oneskunk who stood six-feet-two but weighted only 150 pounds. It states that he was a man of few words. According to the article he seldom said anything but "ugh," which meant "yes," "no," or "maybe," or anything he wanted it to mean. Oneskunk seldom specified. "Even in Britain, where taciturnity is a national trait, Private Oneskunk was a very quiet man."

The fact that he barely spoke was not the only thing that won him the attention of British commando experts who were assisting in training him and other American soldiers who were Rangers.

"There was his name, for instance. American officers explained that Private Oneskunk is a full-blooded Sioux Indian, from Cherry Creek, S. D., and that out in his part of the country, the Oneskunk family is very well known indeed."

The article goes on to state that one day when the rangers went into some heavily-wooded country on maneuvers. It was an exercise in woodmanship, and particularly in traveling through underbrush without making any noise. A British officer thought Private Oneskunk wasn't paying much attention.

An American Officer responded "Well, you see sir, this is just child's play to him. He's probably known how to sneak right up on a chipmunk ever since he was six years old."

So the officers put Private Oneskunk on his own and told him to show them something, which he did.

"He disappeared into the woods, leaving his unit. He was supposed to try to creep back to it without being heard. About ten minutes passed. The whole force, officers and men, strained their ears for some sound of him. They heard nothing.

Then they heard an "ugh." Private Oneskunk was sitting in a tree, right over their heads."

The article goes on "Just to prove it was no accident he went away several more times, and returned, as soundless as a shadow, until he was close enough to touch them. Once he crept up beside one of his mates and whisked away the soldier's rifle. They neither saw nor heard him.

And so, in no time, Private Oneskunk became a full-fledged ranger. Presently, he found himself with his unit, aboard a transport. He said he hoped the ship was bound for India, because he wanted to be in a country full of Indians. They told him those were not his kind of Indians. "They wear beards," they said.

"Ugh" said Private Oneskunk. "Different tribe." He is somewhere in North Africa today.

Miles S. Horn

Another interesting person from the same time period was a man named Miles S. Horn, otherwise known as "White Crow" to his Arikara Indian fellow-tribesmen. An article in The Salem News, Tuesday January 25th, 1944 states that he is known as "Chief" to his Army buddies. It goes on to state that: "he would rather shoot a bow-and-arrow than a Garrand and he can stalk a man as well as a puma.

That's why he's teaching his craft now to soldiers training in commando and ranger tactics.

"Stalking game requires all a man's cunning," the 50-year-old Indian corporal says. "He must know how to hide better than the animal he is after. That's why I want to teach all I know to other soldiers who are learning how to stalk the most dangerous prey."

Miles S. Horn was born on the Fort Berthold Indian reservation near Bismarck, North Dakota and he had a varied career as a trapper and hunter, baseball player, missionary, range-rider in Montana, movie actor and artist. He had been married twice, "once to a beautiful Cheyenne and once to a comely Sioux, and is the father of seven children."

The Billings County Pioneer, November 19, 1959 provides some more background about Horn. It states "Childhood days were carefree days for Indian children. White Crow, who received his name from an uncle, early learned to hunt small game with a bow and arrow. He thrilled at the festivities when the large tent encampments of the tribe brought him new friends. The other children often stopped their games to dash over to see what he had drawn next." Horn was an artist from a young age.

"White Crow was born at Elbowoods, July 15, 1895. His mother's name was White Calf and his father was Izaak Fox. Both were Arikarra Indians. Later White Crow took the name of his stepfather when his mother married Strieby Horn."

The article mentions that Miles Horn's uncle Red Star, was one of the Seventh Cavalry enlisted Indian scouts for General Armstrong Custer. He recalled the story told him by his uncle of the strange atmosphere which circled Fort Abraham Lincoln "when Custer's women said goodbye, as the band played and the men rode forward, never to return, that fateful spring day of May 17, 1876. Red Star was one of the few survivors under Major Reno who escaped to tell his story."

The Salem News article goes on to state that it was the trapping and hunting that developed Horn's stalking prowess. "He always has returned to the woods when other livelihoods failed and the going got tough.

His baseball career was limited to a season at Grafton, N. D., where he played with Walter Johnson the year before the latter joined the Washington Senators.

In his grease-paint days. Horn played in the Fox movie, "The Oregon Trail," with Tom Mix and the as of yet undiscovered Gary Cooper. He was featured as the buffalo herdsman.

Overage for active service now, the Arikara corporal went into the camouflage school at Camp Clairborne, La., last July. He wants to stay in service for the duration, so he can teach the tricks of the Indian trapper to other soldiers."

According to the Billings article "Loyal to his government, White Crow served in World War I and II. He proudly recounts that the second time he volunteered, for special service for men in the 40-50 bracket, he received top rating physically, although half the men did not pass. He took training in camouflage work and later trained others at camps in Illinois and Louisiana. He was discharged from the United States Army with a 70 percent disability incurred while in the line of duty.

Three others in his family saw service with him, making quite a record. His son Denver was with the Marines and his daughter Willena was with the WAACS in World War II and his son Harold took part in the Korean War."

Regarding Horn's career as an artist the Billings article provides some details:

"While roaming through Montana, White Crow struck up a friendship the Artist Charles Russel, who advised him, "Your best teacher is Mother Nature—keep her sunset colors in your paintings." Later he took special work at the Otis Art institute in Los Angeles and met Norman Rockwell in his class, who became his friend.

White Crow's paintings were exhibited at the early American Art Week Exhibits held at Bismarck. By his estimation by the time of the article he had painted over one hundred pictures.

The North Dakota State Historical Society at Bismarck has two of White Crow's paintings on the second floor of the museum. They are of an Indian hunting buffalo on horseback and an Indian village scene along the Missouri river.

Murals by White Crow decorate the Bob Totman Frontier Store, Sheridan, Wyoming and his paintings are displayed at the General Custer hotel and the Stockman Bar at Billings, Montana."

The article also mentions that an oil painting by White Crow was commissioned by Mrs. Franklin D. Roosevelt and once hung in the White house. His painting, "The West's First Americans" won first place in the Wyoming State Fair in 1954, was displayed in the Verihoff Art Galleries in Washington, DC. And included in the United States arts and crafts display at the 1958 World's Fair in Brussels, Belgium.

This is the copy of the painting which Mrs. Franklin D. Roosevelt asked White Crow to paint for her of the way the Indians used to live. Mrs. Roosevelt hung the painting in the White House and later it was included in the Hyde Park collection. White Crow learned on a recent visit to Washington, D. C.

Ernest E. McClish

An article in the Okmulgee Daily Times, April 26th 1945 discusses the exploits of Ernest E. McClish who was a Choctaw from Oklahoma who graduated from Haskell Institute and Bacone College, and was called to active duty in the National Guard in 1940 and sent to the Philippines.

The article written by Elizabeth Kirk states:

Lt. Col. Ernest E. McClish, world famous guerrilla leader of the Philippines, is spending a leave with his mother, Mrs. R. E. McClish, 1211 West Eight. He sailed for the Philippines about four years ago to train the Filipinos to fight and became a guerilla leader after the fall of Bataan and Corregidor. His wife, daughter, and 3 ½ year old son, whom he had never seen before his recent arrival in the states, are with him.

"And the invading Yanks were aided by Filipino guerrillas."

This single paragraph is found in reports of nearly all the Philippine Islands thus far. Behind such statements lies many dramatic stories, similar to that of Lt. Col. E. E. McClish, one of the earliest organizers of the roving, swashbuckling Philippine guerrillas, who led guerilla units on the island of Mindanoa from August, 1942, until his recent return to the Sates.

When the Japanese invaded the island of Mindanoa on April 29, 1942, Col. McClish was stationed at Malabang, Mindanao, the same place the American forces entered in their recent return to the island. "My unit came in contact with the enemy before daybreak the following day. I was in command of the Third Battalion, 61st Infantry, and being the senior officer at Malabang, Col. Mitchell placed all other units at that station under my command," said Col. McClish.

"After fighting for four days and nights I developed malaria and was taken to my division hospital at Dansalon, Lanao, and from there hospital patients were evacuated farther into the interior of the island because of the enemy's rapid advance," said the colonel. "As a result the bed patients where I was located were unable to surrender when the 81st (Philippine Army) Division surrendered on May 18, 1942." Col. McClish escaped capture although the Japanese put out a bulletin announcing that unless all Americans surrendered by June, 1942, they would be hunted down like dogs.

Col. McClish started organizing pre-surrender soldiers into a combat organization about Aug. 1, 1942. "By September '42 I had an organization of over 300 soldiers and had collected four 30-caliber machine guns, over 150 rifles and six boxes of 30-caliber ammunition," said the colonel in describing his start as one of the earliest organizers of Filipino guerrillas.

About this time guerrilla bands were starting all over the eastern portion of the island and while organizing the eastern province Col. McClish received word of a certain Gen. Fertig who had started an organization on the western part of the island. "I thought I had better contact Fertig before going further so I took a small sailboat across the sea and arrived at his headquarters about Nov. 15," he said.

Gen. Fertig was a colonel in the Engineer's Corps AUS. His command and that of McClish were consolidated and McClish was given the mission of organizing the eastern provinces of Mindanao, consisting of Oriental Misamis, Agusan, Surigao and Davao, and his command was designated as the 110th Division USFIP.

"The task of unifying the roving, some irascible, others committing depredations amounting to banditry, independent guerrilla bands in the four big provinces under my command, was rather a difficult problem in the beginning," said the colonel. They had to be molded into an army of compact, disciplined fighting men capable of stalemating, harassing and confining local enemy garrisons.

"Some of the guerrilla units showed indifference if not an attitude of hostility," says McClish. "Most of them were lacking in military training and their leaders were drunk with their newly acquired powers and reluctant to submit to a higher command and a few of them showed an anti-American attitude. These were disarmed and jailed and their followers absorbed into the new organization."

The friendly understanding Filipinos in these units were given due consideration and were commissioned, others promoted and allowed commands. The ones in between the hostile and the friendly were won over by kindly and more persuasive means.

McClish had with him 14 American officers and about 70 enlisted men in that area who had escaped capture or who had escaped from Nip concentration camps. "Filipinos do much better under American officers and have profound respect for Americans in general," said the colonel.

In Col. McClish's command area a Philippine army division was formed in accordance to the Philippine Army Tables of Organization with full strength regiment in each of the four provinces.

"Then there was the task of organizing the different regiments. Despite lack of transportation and communications facilities, within a few months or before March, 1943, three regiments were fully organized and another started, and before this time the division headquarters staff offices had been filled in."

In each of the four provinces civil or provincial governments were formed with the organizations of the regiments.

The colonel's headquarters was near Buenavista in the Agusan River valley. The position was chosen because of the river down which it was hard for the Japs to travel because the trails were extremely rough. The Japanese attacked my headquarters every day with about 15 planes strafing and bombing," said the colonel. "They came in in March '44 and it took until July to occupy the Agusan River valley and prepare for the American invasion."

Col. McClish was given much publicity in Japanese newspapers on the island and 15,000 pesos were offered for his capture. "Once I was even reported dead," said the colonel.

McClish's guerrillas had nine launches, four of which were captured from the Japanese, and four motor sailboats. "We used our launches in the sea more than the Japanese although they controlled the water surrounding the island," said the colonel.

During the time they functioned from Sept. 5, 1942, to the end of the year 1944, these guerrillas had 350 encounters with the Japanese, killing over 3,000 of them.

"The Japanese army of occupations was made up largely of Japanese civilians already in the islands, and the guerrillas kept them confined to their garrisons. There were Philippine agents inside the garrisons and in our area the guerrillas had superiority of numbers," said the colonel.

"One of our most interesting battles with the Japanese was planned on my birthday, Feb. 22, 1943," said the colonel. "The 110th moved to my headquarters and on the next day a group of a little less than 2,000 moved by sailboat to Buenavista, 17 miles away, and made plans for an attack on Butuan.

'A column seven miles long started the march from Buenavista with bands leading the battalions and carrying Philippine and American flags. Civilians lined up along the way and they furnished food for the soldiers," said McClish.

"On March 3, early in the morning, Butuan was attacked from three different directions. The fight lasted nine days before the Japanese garrison was taken over. Three launches were captured and a variety of shoes, clothing, canned goods and 103,000 pesos in money."

Describing the cruelty and barbarism of the Japanese, Col. McClish gives an example of the treatment of two officers and five enlisted men who came from Leyte to Mindanao in sailboats. They were captured by the Japanese, who tied them to trees.

"For the first three days they were given cocoanut water and no food. The following four days they were given nothing. The two officers were tortured to death and the enlisted men bayoneted after they had dug their own graves," Col. McClish related.

"The Japanese covered the graves with very little dirt and shortly after some Filipinos recovered the bodies and found two of the bayoneted men still breathing. They cared for these men until a doctor could be found. Both lived although thy had been bayoneted many times through the abdomen."

Describing their recovery as short of miracle the colonel said that the physician said, neither of the men would have survived had they been given any food.

"The bayonets had slid past the intestines and they were not punctured although the points went through their bodies."

"The Filipinos have learned a lot during the war toward doing things for themselves. I always had the closest confidence in the people as well as the civil government. They appreciated whatever measures were taken for the benefit of the majority. And they'll come out on top." Said Col. McClish.

· Major Ernest E. McClish, above, son of Mrs. R. E. McClish, 1211 West Eighth, has been reported by the war department as missing in the Philippines, his mother learned yesterday from Major McClish's wife who lives in Cedar Rapids, Iowa.

A captain in the army reserve corps, Major McClish was ordered to active duty in 1940 at Fort Huachuca, Ariz., and received orders for foreign service less than a month later.

He sailed for Manila Jan. 18, 1941, and was stationed at Fort McKinley to train native troops.

Later he was transferred from Fort McKinley to Panay island, where he assumed the duties of a major.

Last word his mother had from him was a cable Jan. 27.

Bushmasters

Crack Indian jungle fighters: Trained as "bushmasters," members of 20 American Indian tribes have helped smash advanced Japanese positions in South Pacific Islands. Shown here are, left to right, Pte First Class Dale Winney, Pte Joe Tapaha, Pte First Class Joe Fishi, and Pte Perry Toney, all from the US South-West. Winney, Tapaha, and Toney are wearing regalia they made for celebration of their traditional Christmas dances.

On June 21st 1942 for The Knoxville News Sentinel in an article titled U.S. Training New Kind of Soldier for Warfare in Jungle, Nat A. Barrows writes the following:

In the Jungle, Panama... Out here in the eternal mud and dampness of the great shadows, a new kind of American fighting soldier is living an adventure tale of cowboys and Indians that makes the usual Wild West yarn sound like the exploits of the late George Apley at a Boston high-tea.

Never before has the United States had soldiers or a type of warfare such as this correspondent has seen here in the jungle.

Utilizing the skill and background of honest-to-goodness cowboys and Indians from the Southwest, along with a sprinkling of rugged lads from the cities, the Army has developed an Infantry patrol method for intensified guerrilla fighting. It has studied the tricks of Dan'l Boone, Natty Bumpoo the Deerslayer, John Brown, the British Commandos and the Japs, and it has taken something useful form them all.

Emerges: The Bushmaster.

As deadly and as dangerous at the exceedingly venomous jungle snake from which they get their name, these Bushmasters of Panama prowl the uncivilized trails as advance guards, reconnaissance troops, intelligence agents, jungle commandos, hand-to-hand fighters adept with terrific modern firepower or the primitive machete.

They're always ready to move at once—and to move through strange terrain silently and quickly. They know how to set an ambush in the jungle. They know all the deadliest tricks of jiu-jitsu, all the head-chopping thrusts of the machete, all the ways to live off the land and still strike hard in close-range combat.

These Bushmasters are trained to cover unexplored terrain, swim treacherous rivers and scale jungle peaks so efficiently that the average soldier could not keep up with them. They have special, carefully tested "iron rations" sufficient for many days; they have their own uniform and equipment, they have their own thoroughly unbreakable code of communications.

Almost all of these hundreds of Bushmaster soldiers speak Spanish. This is an important factor in their visits to little jungle villages where Americans never were seen before. But, more important in wartime, they are able to send out their radio messages by walkie-talkie sets in the Indian dialects of the Pimas, the Papagos, the Apaches or the Navajos. Or even in two or three Chinese dialects, if they wish.

Their commanding officer, Col. J. Prugh Herndon, of Tucson, Ariz., to whom the Indians are as devoted as to a favorite uncle, merely dictates in English a field message which he desires to clear into another area. One of his Indian Bushmasters quickly recites it in his own tongue to another Indian walkie-talkie operator. The message is as secretly transmitted as if it were in a four-hour code.

On the trail, as they prowl silently, well-spaced for instant hand-to-hand combat with the enemy, the Bushmasters have their own specialized communications. An upraised hand... a twist of the head.... A tap on a rifle barrel... an imitation of a certain jungle animal or bird. Surprise is a basic element of their work—and silence is the key.

Roaming the jungle with the Bushmasters, I have seen them disappear into the stifling, humid air by clever use of the vegetation. They were hardly two yards away but I could not see them although I'd watched them drop, helmet fringed with reeds, bodies wrapped with banana fronds. Imagine their invisibility to an enemy trail party walking unwittingly into such a Bushmaster ambush.

I've seen them hack their way through jungle brush where the sun never had shown, flicking their wrists so that the machete blade cut a path with a minimum of muscle energy. I've seen them plunge into rivers amid alligators and crocodiles, waterproof packs floating behind, small balloons keeping ammunition and cigarettes dry, rifles held aloft for instant action. I've seen them practice "boobie traps" for the extermination of enemy guerillas, and show how to break the neck of any opponent foolish enough to wrestle with them, and reveal the old cowboy and Indian tricks of making a camp wherever night found them, unmindful of pelting rain and knee-deep mud.

"These men play for keeps," said Col. Herndon, admiringly. The colonel was raised in the saddle as the son of a missionary to the Indians. He speaks their dialects and he knows their psychology. He speaks fluently the language of his American-Mexicans of whom there are scores among the Bushmasters. The Mexicans too, make excellent jungle patrol soldiers.

There are Midwesterners and Eastern city boys out here with the Bushmasters too. Lean, eager young men from Ohio and New York and the industrial cities. What they lack in childhood background as cowpunchers and ranch-hands they make up in enthusiasm to learn. All are volunteers, and all are hand-picked for physical and mental ability, native intelligence and adaptability.

They've got to have what it takes or they won't last long in this dangerous and highly adventurous new kind of infantry warfare.

Take a few of these Bushmasters at random and follow them through the jungle. Take for instance, Sgt. Jesus Cortez, a Yuma, Ariz, Mexican; Pfc. James Gardner, a real Navajo chief; Sgt. Gordon Hawkins, one-time cowhand from Douglas, Ariz, and such ex-tenderfeet as Pvt. Edward Preleyko of Cleveland, O.; Pvt. George A. Squires of Butler, Pa; Pvt. Albert Hathaway of Cincinnati, O.; Corp. Robert Slater of Preoria, Ill.; Corp. Kendall Upchurch, of Denver, Colo., or Pvt. Sherman Daugherty, of Norwalk, O.

The terrain problem is the crossing of a river. Capt. Nelson F. Huie, of Phoenix, Ariz, and Lt. Erwin R. Bennett, of Mangum, Okla., merely give hand signals indicating the course. Quickly the Bushmasters squat down and begin rubbing mud across their faces and necks. They are coming out from the shadows; the shine of their faces must not give an airplane observer a hint as to their presence.

Packs are adjusted, and down they go, as silently. As the deadly coral snake which Capt. Huie has just killed—right under my feet. They swim with one hand, hold their tommyguns, automatic rifles and carbines clear with the other. The balloons float on either side of the packs.

They make the other side and scramble slowly up the sharp bank. Their movements are deliberate and their muddy faces are down. In that way lies safety from air detection. In 10 seconds they are swallowed by the jungle. Only the rustle of jungle animals cuts through the depressing still heat.

Col. Hernden watches from the other bank. At his command, the Bushmasters pop up from unexpected hiding places and again cross the river. They scramble up like mountain goats—or the Bushmaster snakes.

"These men of mine are rarin' to go—anytime, anywhere," says Col. Herndon. And he smiles a great smile of delight, even though he knows that many of these Bushmasters of his soon will be absorbed into other regiments here in Panama and he must start training more doughboys to fill their places.

U. S. Troops Duel with Machetes

American troops in the Panama Canal Zone have been undergoing intensive training in all kinds of jungle fighting. Included in the courses is the proper use of deadly machetes, big native knives. Pvt. Roy L. Barnard (left), Sand Springs, Okla., and Sergt. Gordon Hawkins, Douglas, Ariz., are shown practicing duelling with the sharp weapons. Although Barnard has fallen (top) the fight is not over; he is not as helpless as he looks. He recovers (bottom) and slashes away at Hawkins behind the knee. — (Central Press)

Brummett Echohawk

In Brummett Echohawk: *Chaticks-si-chaticks* by Kristin M. Youngbull she notes that Brummett Echohawk felt a strong sense of patriotism as an American, and a strong connection to his Pawnee heritage. "He felt no need to choose between the two. By serving honorably in the U.S. Army he and his brothers could become warriors among the Pawnee people. He saw value in his family and tribal martial traditions, and felt that American soldiers had much to contribute to the military."

She goes on to write that "Stereotypes of American Indians as warriors became rampant during the World War II era, and the media used men like Brummett Echohawk who excelled in their military training to promote the stereotype, and to garner support for the war. His participation, along with numerous friends in the New York parade and Major Gilstrap's description of him as a hand-to-hand expert in the nationally syndicated article about the virtues of American Indian soldiers in the U.S. Army offer examples of how the media drew him into the larger phenomenon generated by government officials and the press. Generally he embraced the idea of American Indians as exceptional warriors. He willingly went out on long-range patrols behind enemy lines, and played on German fears of Indian soldiers."

Youngbull mentions that Echohawk accepted that as an individual he served as a representative of American Indians. Although he served in a predominately Indian unit, per United States policy, he also served among non-Indian soldiers. He once said "Whether you know it or not, you as an Indian are an image. You always will be."

Echohawk's uncle George used to explain to his boys "that going to war meant that they could come home and hold their heads up. No one could talk down to them, or speak poorly of them, because they had proven themselves by going in battle."

She writes that Echohawk "also had the opportunity to help show the rest of the nation, and Europe, what American Indians could do – what a grandson of the Army Scout, Echo Hawk, could do. He had an intense desire to live up to the family name."

As a young soldier, Echohawk could saw his personal connection to Pawnee and family history. His grandfather and father had worn the uniforms of the United States. "The Pawnee Scouts had served as part of the 45th Division during the conflict with Pancho Villa. Patton had also served with the Pawnee Scouts in the expedition against Pancho Villa, and then spoke to Echohawk and his friends of the value of their service in Sicily. He and other Pawnee soldiers joined a warrior society brought to life by Pawnee soldiers in World War I. Echohawk also made himself a part of the history which unfolded in World War II through his fighting efforts, and also through his artwork.

Anzio Advance pictured by Fighting Artist

This is how war looks and feels to an American infantryman in Italy—mostly mud and death and drudgery. Sgt. Brummett Echohawk sketched this scene as his platoon slipped and slopped ahead toward Carrocetta, one of the hottest spots on the Anzio beachhead. At lower right is a dead German, a gas mask cannister beside him. The U. S. soldiers carry battle packs.

This war's first actual battle sketches to be drawn by a front-line infantryman are those of Sgt. Brummett Echohawk, pictured above. A full-blooded Indian from Pawnee, Okla., this 22-year old soldier and artist has been in the Army more than two years and has been wounded twice—at Venafro and on the Anzio beachhead. While convalescing, he finished a number of rough sketches which were made under fire.

Having trained for nearly three years in the United States as part of the 45th Infantry Division, and made it through North Africa, Sicily, and southern Italy, Echohawk became a seasoned soldier. Although he had been injured numerous times in battle, "the damage done thus far proved insufficient to make him leave his unit on the front lines. Severe concussion affected his hearing and caused internal damage, but he persisted. His sense of duty, his determination to prove himself a true warrior, his camaraderie with Indian and other soldiers, and his faith in the Creator, came together to bolster him throughout the war. All the while, he continued sketching—making a record of what transpired."

Youngbull writes that as a soldier, Echohawk calculated risk and took some chances. "He once explained to a friend that he may have calculated risks somewhat differently than others because of a certain faith he had in a sort of Pawnee tradition. Once the men of the 45th Division knew they were going to war, Echohawk and some friends apparently visited one of their tribe's spiritual figures. The individual told them that of the group, one would be killed in war. The others would return home."

One of the friends died almost immediately after landing in Sicily. Echohawk understood the death of his friend as an indication that he would make it home. According to Turnbull, Michael Gonzales the Curator of the 45th Infantry Division Museum in an interview on May 8th 2012 stated "In Brummet's mind, this was a pass. 'Okay, one of the four of us is gone; I'm going to go home. I'm not going to be killed in this war. I can do anything I want and get away with it.'"

Echohawk later realized that he had not been promised that he "wouldn't get all shot to pieces, only that he would not be killed."

Turnbull writes "Add to the experience with his friends, the Pawnee legend that "one Pawnee warrior will die on the Warpath for every major Pawnee battle." Echohawk referred back to the experiences of the Pawnee Scouts and Pollock who died of illness related to his service in the Spanish-American War. He noted that in the Mexican Border Expedition, the Pawnees did *not* lose one man but suggested that this resulted from the fact that they did "no real fighting there. In World War I, a Pawnee died in action from poison gas."

Echohawk did not view the One Man legend as a negative or a curse. "He compared it to a large tree falling in a forest to provide for saplings. He wrote: "Death is a part of creation. One life goes that another may live. The Old Ones tell us this… No life is really wasted. We accept this."

He took chances on occasion because he believed that he could make a difference without getting killed. "He fully intended to live up to the standards of *Chaticks-Si-Chaticks*, or Men of Men. He said: "In the modern world we sort of joked about it—who would be the man? All of us were hit time and again, and we lost one man."

Echohawk and his comrades told a story about a night patrol through German lines. They had made it to their objective without incident. ON their return, they stopped short of their lines for a brief rest. Feeling that they were comfortably close to their own positions, they put their weapons down and relaxed against a stone wall.

"Suddenly, the men heard voices speaking German coming along the other side of the wall as a German patrol returned to their lines along the same route. Most of the group had the automatic impulse to lay still and silent and wait for the Germans to pass because the enemy had caught them off guard and they did have their weapons read. "That wouldn't' have suited Brummett one little bit," said his friend, Michael Gonzales, "so [Echohawk] leaps up, and he whips this Bowie knife out, and he lets out a war whoop."

Gonzales continues "He leaps over this wall and charges into the Germans. They scattered like roaches. He keyed on this one German soldier who's

135

carrying a machine gun. Machine guns are heavy. He figured this would slow this guy down. They're running down a slight incline, and the guy's got this machine gun over his shoulder as he's running, and he glances over his shoulder, and he can see Brummett, who, well this is a wartime photograph of Brummett as you can see he's a Hollywood Native American, okay? Put war paint on him and a feather in his hair and he looks like he came out of Hollywood. Well, that's what this German saw. He saw a read Indian chasing him. He threw the machine gun down to get better speed. Brummett caught him anyway and dispatched him post haste."

Turnbull writes: "He then ran back to the wall where his men still stood. Many of the startled Germans had dropped equipment near the wall before beating a hasty retreat. Echohawk's sudden action had surprised his own men too. In the time that it took for them to register what had happened, stand up, and ready their weapons, most of the Germans had exited the area, and Echohawk had already embarked on his hot pursuit of the German fleeing down the hill. One of his friends later told Gonzales that Echohawk's actions that night "scared the hell out of me." He went on to say: "I don't know why. I should have known Brummett was going to do it." Once again, Echohawk caused his friends to wonder why he would take such risks, and once again, his belief that he would make it home had influence his decisions."

The story reminds me of what James Black wrote:

It is likewise their business at all times to endeavor to annoy their enemy, and save their own men, and therefore ought never to bring on an attack without considerable advantage, or without what appeared to them the sure prospect of victory, and that with the loss of few men: and if at any time they should be mistaken in this, and are like to lose many men by gaining the victory, it is their duty to retreat, and wait for a better opportunity of defeating their enemy, without danger of losing so many men.

Echohawk seemed pretty sure that his actions were not going to negatively impact his fellow soldiers. He succeeded in harassing the enemy and him and his men lived to see another day.

In the Blackwell Journal Tribune, January 26th 1962 in the Hunting and Fishing column, Jack Wilkinson writes "Echohawk, syndicated cartoonist and illustrator and descendant of the famed Pawnee Scouts tribe, is faster than most bowmen because he shoots from the hip. And his aim is deadly."

Echohawk tells Wilkinson "When I am in practice, I can split two of three arrows I shoot at. Last summer the best I could do was 12 of 18."

Echohawks' wife, Mary, sometimes helped in the exercises. He tells Wilkinson "Last summer she hung a spool by some thread and started it swinging and I could shoot an arrow through it."

Wilkinson writes "Echohawk, a deep-chested, 190 pounder, is earnestly training for the big buffalo hunt slated for an area near Atlanta, Kansas, during which all participants will be on horseback. As part of his training to strengthen his arms so he can use his powerful bow better, Echohawk walks on his hands for 50 yards each day. The neighbors wondered at first, but now they're used to it."

Forward Under Fire

The entire area was under intense fire when Sgt. Brummett Echohawk sketched this little group of veterans as they prepared to follow a rolling barrage at the Mussolini Canal in Italy. They were tired and dirty after fighting in the Carrocetta region of the beachhead for more than a month. The soldier in foreground is attaching to his rifle the Army's newly issued short bayonet; the corporal in rear carries a rifle grenade thrower. Sgt. Echohawk, an Indian from Pawnee, Okla., finished this sketch in a hospital after he was wounded.

Kenneth Scissons

Dubbed the "Mustachio Commando" by the Nazis because he always wore a handlebar mustache, Sgt. Kenneth Cuthbert Scissons had his exploits depicted in a comic book strip. According to an article in South Dakota Magazine, Scissons was a Lakota-English-Norwegian soldier who became a one-man army during World War II.

Scissons was born in 1915 on a ranch south of Colome, "but came of age in Rapid City, where his father, John, worked for the Warren-Lamb Lumber Company. His Lakota lineage stretched back to his grandmother,

Hannah Mule, twin sister of Little Big Man, the renowned Ogala Shirt Wearer from Crazy Horse's band."

According to the article Scissons was picked on while he attended Rapid City Indian School through the sixth grade. He proved to be a natural athlete who excelled at every sport, including boxing. In the seventh grade he was transferred to public school. During his senior year he got into an argument with his basketball coach and dropped out.

For a while Scissons worked at Warrant-Lamb, spending 10 hours a day shoveling sawdust into boxcars. He was thankful for the opportunity and later recalled "Times were tough. If you ever stopped and put your shovel down somebody else would pick it up, then you no longer had a job."

During the Depression when Warren-Lamb cut back, Scissons joined the Civilian Conservation Corps (CCC) and was posted to a camp near Hill City. "He stood a stout 6 feet by then, and had energy to burn. At day's end, when the other workers clambered into trucks for a ride down the winding mountain road to their base camp, Scissons would set off cross-country, racing over the rugged mountain on foot."

Scissons' oldest daughter Ruth Ahl said, "He ran so fast he thought his heart was going to burst, and he was always at the camp, leaning against a tree smoking a cigarette when they got there."

Scissons eventually married and joined the South Dakota National Guard in 1936. "After basic training he was assigned to the headquarters company of the 109th Engineering Regiment in Rapid City."

According to the article he "signed up for a second hitch three years later, a perilous moment in history. Germany overran Poland in the fall of 1939, plunging Europe into a war that many American's feared would soon be at their doorsteps. Scissons and every other volunteer surely understood that the Guard might soon require more of them than one weekend a month, but they signed up anyway."

Sharon Schaefer, Ahl's younger sister stated "Dad was extremely patriotic. He believed in doing whatever was necessary to protect the country. That was how he was raised. You were loyal."

The article mentions that Scissons and five other soldiers from the 109th volunteered for federal service and were shipped overseas almost a year before the attack on Pearl Harbor. "They were among a contingent of Americans who trained with elite British Commandos and took part in raids on occupied Europe, the idea being that they would rejoin their old units when U.S. forces formally entered the war. Their combat experience would season the mass of green troops."

An article in the Rapid City Journal, 22nd April 1943 states that he "was in the National Guard for about six years and served on the state staff at local headquarters from Oct, 10, 1940 until he left, Dec 20, the same year. He sailed Jan 14, 1941, and was with the first bunch to land in Ireland. Later he went to Dundee, Scotland for British Commando training."

The article continues to state: "That led to his training as a Commando, or Ranger, as the American Commandos are called, but later his unit was depleted from hard service and was broken up. He returned to his old regiment but reported that he found "patrol pretty dead" compared with his previous excitement and prowlings.

The World War II hero received his Indian blood from his father's side. The grandfather, John. C. Scissons, was born in England, but by 1876 was freighting from Sidney, Neb., to Deadwood. He often related that early in August 1876 he and his freightors were camped near Buffalo Gap when a man calling himself Jack McCAll appeared and claimed that he had shot and killed Wild Bill Hickock in a saloon in Deadwood that day, Aug 2.

Scissons and the others scoffed at him, but upon their arrival in Deadwood a few days later they discovered it was true. Sometime afterwards McCall was hanged for the killing.

Scissons during his freighting days met and married a Pine Ridge Indian woman who became the present John Scissons' mother and Kenneth's grandmother. The father of the present Mrs. John Scissons was born in Norway."

When the Allies invaded North Africa on November 8, 1942, Scissons was put ashore on Algiers, Tunisia. An article in the Rhinelander Daily News, January 6th 1943 discusses a major incident which Scissons was involved in during the war:

"The leading scorer in the race by commando-trained American soldiers to see who can kill the most Germans is a young South Dakota lumbermill worker of Sioux Indian descent, who has 10 notches on his Garand rifle."

The article continues to state that Scissons: "is not engaged in patrol work, in what he calls the "sissy size" Tunisian hills.

His record string was chalked up in less than four minutes during a sortie by British-trained units near Bizerte last month.

This daring engagement behind enemy lines was one of the bloodiest in which American troops, trained in the hit-and-run tactics of the British commandos, have yet participated.

Before dawn, he and others of his unit were on the coast so close to Bizerte they could make out the airport. The object was chiefly to harass the enemy behind his own lines and in this the men certainly succeeded.

After reaching their objective following an all-night 15-mile march. Scissons and some others were ambushed by the Germans in a hilly area.

"They pinned us down with tommy-gun fire from a ridge," said the sergeant. "I told our leader our only chance would be for some of us to go around the hill and clean them out. I volunteered to take my subsection.

The only way to get up was to follow the creek bed for 150 yards, but we had to run through open country to reach it. I didn't stop to look behind but when we reached the creek bed only five of our original 12 were left.

We got halfway up when we ran into more tommy-gun fire and lost another man. As we were than too few to hope to take the ridge we decided to withdraw. We thought the best plan was for two of us to make a break for safety while the others protected with covering fire as well as possible.

Guy Wright, of Henryette, Okla., and I stayed behind when our other two fellows started back down the hill, hitting the ground about every 10 feet like jackrabbits. Some Arabs, or Germans dressed like Arabs, shouted and began pointing. Hidden German soldiers then raised up to fire.

That was the first good glimpse of them we got and we really went to town on them. Wright, whose buddy had been killed only a few minutes before, threw a hand grenade at the nearest German. It landed right at his feet and he went in a dozen different directions at once.

I had begun to clear off our section of the ridge. It was like popping off squirrels, only I was never able to get anywhere near that many squirrels in anything like that time. You can't fool 10 squirrels in four minutes.

The Germans were about 10 yards away, except for the one that Wright picked off with his grenade. The whole thing was over in less than four minutes and you know they never even scratched our decoys."

An April 16th 1943 article titled "Sturgis Soldier Acted As Decoy For Indian Hero" in the Deadwood Pioneer describes the incident this way:

"The fabulous exploit of Sgt. Kenneth Scissons of Rapid City in killing 10 Germans during a four-minute engagement near Bizerte in North Africa in December, was made possible because another Black Hills soldier acted as his decoy.

The "decoy" was Jerry Gorman, son of Mr. and Mrs. Andrew Gorman of Sturgis, who has two notches on his own gun. He modestly revealed his part in the affair in a letter written March 17:

"'You probably saw that write-up in the Rapid City paper about Kenneth Scissons on the African front—I was with him—I was his decoy that he mentioned in the paper—We had quite an experience—Will tell you about it sometime."

The article mentions: "Scissons related to an Associated Press correspondent who termed him the leading scorer in the race among American-trained Commandos to kill the most Germans, that while on a sortie behind the enemy's lines, his unit was ambushed by Germans in a hilly area.

Seeing they were trapped, Scissons and 11 other men volunteered to clean out the Germans.

Seven of the 12 were killed in a dash up a stream toward the hill on which the Germans were hidden. However, the remaining five started up the hill. Halfway up, they ran into fire and another man was lost.

The last four, thinking they were too few to wipe out the nest, decided to withdraw. Gorman and a companion, it is now known by piecing together the Sturgis soldier's revelation with Scisson's story, started back down the hill while Scissons and his companion covered them.

The fleeing men drew the fire of the Germans, who raised up from their nest, exposing themselves to Scissons and his buddy. While bullets were whizzing around Gorman and his companion, Scissons and his buddy blasted away. During the ensuing four minutes, the Rapid City Indian accounted for 10 Nazis."

The South Dakota Magazine article states that Scissons earned four Bronze Stars and a Purple Heart, in addition to the Distinguished Service Cross, during his time in service. It also mentions that he was a good soldier in every respect but military discipline "a shortcoming that could be attributed to his pugnacious character, the very quality that made him an exceptional warrior.

A man who served with him in Tunisia, Chuck Cory stated that "Kenny was a one-man army over there, he'd leave base and we'd never know when he was going to come back"

According to the article "Scissons infiltrated enemy outposts, struck silently, and then left a calling card to demoralize the rest. The German's paid Scissons the ultimate compliment by hanging wanted posters of him across Tunisia."

When the Allies landed at Anzio, Scissons was asked to penetrate enemy lines and capture Germans for interrogation, "an assignment he called his worst of the war. Month after month of mortal danger and operating on his own left him with little patience for rear echelon types, even those who outranked him."

Scissons returned from the War in the spring of 1945. His first stop wasn't home but a base in Texas. His daughter Ahl said "He was a trained killer, so they thought they had to mellow him out, I guess… you don't just come right back to your family after something like that."

After the war Scissons entered law enforcement as a conservation officer with the South Dakota Game, Fish and Parks Department. After twenty years he finished his career as a criminal investigator with the Bureau of Indian Affairs. Scissons died on September 9, 1973.

Indian Scouts

An article in the Richmond Times Dispatch, April 4th 1942 describes U.S. Army Indian Scouts who are on active duty. The article has the above photo and states "The weather-beaten features of three members of the United States Army Indian scout organization at Fort Huachuca, Arizona. They are (left to right) Corporal Jim Lane, John Rope and Kassey &-32. Only seven Indian scouts are in active service now. Sergeant Sinew L. Riley of Indian scouts plays with his son, 7-month-old Haystack, who is strapped to Mrs. Riley's back.

The article continues "Privates William Major and Andrew Paxson are on patrol duty along the southern border of the Huachuca Mountains. Indian scouts originally were used by the army after the Civil War. When the last seven retire and are discharged, Indian scout enlistment will come to an end." (Picture Below)

There were several other photos which were taken along with the photos used in the article above. They were taken by Carl Gaston for the Army Signal Corps on April 1st 1942.

Another article in The Freedom Call, May 14th 1942 states "These famous Indian scouts stationed at Ft. Huachuca, Arizona, were originally mustered into army units shortly after the Civil war. The government no longer recruits these scouts as there are only seven left. When they retire and are discharged this once famous organization will be only a memory. Today the members of the detachment are performing valuable services as reservation range and fire guards. Many tales have been told of the bravery of Indian scouts attached to regular army units.

Corporal Jim Lane, U.S. army scout, second from right, shakes hands with John Rope (Black Larriet) retired U.S. army scout, who tells of the many battles in which he has fought.

Private Andrew Paxson is shown scaling a dangerous peak for a better look-out.

These Indian scouts are filing up the mountainside looking for anything that looks like trouble, for brush fires, and so on.

Sgt. Sinew L. Riley is receiving reports on the activities of the day from his scouts.

Sgt. Sinew L. Riley typifies the eyes and ears of Indian scouts in this century.

Corporal Jim Lane is show here after having quenched his thirst from a spring.

Sergeant Sinew L. Riley is teaching his son, Larrie H. Indian wood lore.

Private Andrew Paxson is shown leaving his Army tent on out post to start his scout duties.

Grey Otter's Knife Act

For whatever it's worth, my hope is that I am always able to convey as much about the person as well as what they taught; and that I am able to give them a little bit more exposure then what they have had in life. Sometimes their stories get lost with the passage of time.

A lot of savvy people already know the Grey Otter story. The one about him taking out the Nazi sentry. They know a few things about him but they don't know the whole story, most of what is known about him today comes from McEvoy's book.

On July 29th, 1963 in an article written for the Los Angeles Times by Art Berman., titled Knife Act Honed, Tough Job Cut Out for His Sharp Eye; discusses Grey Otter's knife throwing act with his assistant.

"Mrs. John J. Anthony Jr., 33, a Tarzana widow with three small children, plans to hold her family together by risking her neck."

Berman writes that: "She has agreed to become a target for a tomahawk and knife thrower who says the only people he's ever hit with knives were enemy soldiers in World War II. 'I've never had an accident,' said Skeeter Vaughan, 40, a half-Cherokee Indian and chief of the Federated Indian Tribes.

"'I know he can't hurt me," said 4ft. 10 in. Little Fawn Anthony, an Apache Indian, who say if he hit her with one of those tomahawks, it would all be over.

Mrs. Anthony, 18617 Clerk St., Tarzana and Vaughan 17840 Willard St., Reseda, an estimator for a concrete firm, hope eventually to get their act booked.

Vaughan who has been throwing knives and tomahawks for years, uses a set of steel tomahawks that were made by an Indian named Lone Horse, "an Osage friend."

Mrs. Anthony, who was married to the estranged son of radio human relations counselor John J. Anthony… became a widow on July 2 when her husband died of cancer at UCLA Medical Center.

She said her husband, a 35-year old artist, left only $20, and she is unable to turn to her father-in-law, now a resident of Beverly Hills, because of a long-standing family rift."

Berman writes "The widowed Mrs. Anthony Jr. said her husband's unexpected death forced her to make a quick decision. 'I'd been a housewife so long, I didn't know what to do,' she explained, 'I had been a target for Skeeter Vaughan at Indian meetings, so we discussed it and decided to make a career of it."

In the article Grey Otter says he feels "perfectly calm" about tomahawk throwing and he estimated "that the 1 ½ lb. weapon travels 75 m.p.h. to 80 m.p.h. during its 20 to 25 ft. journey. He throws a dozen tomahawks forming an outline about 1 ½ in. from Mrs. Anthony's body."

The article states that "He throws knives about ¾ in. from his target and performs other feats of accuracy.

A sergeant in the 18th Cavalry in Europe, Vaughan said he worked behind enemy lines and was forced to throw knives at "five or six German soldiers. "Those were the only casualties I ever caused," he added."

The article continues: "Mrs. Anthony said she's looking forward to her knew career. "I have complete confidence in Skeeter, and it will enable me to keep my children."

In the Los Angeles Times on November 6th of that same year. Things were not meant to be. In an article titled "Thief Halts Tomahawk Act" it states: "A tomahawk-throwing act has been thrown for a loss because someone stole their tomahawks, they complained Tuesday.

Little Fawn Anthony, 33, who is the target for tomahawks and knives thrown by Skeeter Vaughan, 40, said 11 hand-made tomahawks and 18 throwing knives disappeared from their car while it was parked in a Chatsworth bowling alley Sunday night.

"We just got an audition for the Steve Allen Show," said Mrs. Anthony… "We thought it was going to be our big break. But if they call us and we don't have equipment, we've had it."

Vaughan told a Policeman that the knives and tomahawks were worth $600 but they could not be duplicated. Mrs. Anthony offered a $50 reward for the return of the equipment "no questions asked."

Grey Otter, Later Years

In the Valley News, August 16th 1968 there is an article titled Indian Couple Wed During Rites at State Exposition. At the California Exposition in Sacramento the article states that "Grey Otter, champion Cherokee tomahawk thrower, married pretty, blue-eyed Becky Little Eagle in a ceremony conducted on Frontier Street." Both had appeared there with the Dick Shane Wild West Stunt Show.

The article states that the couple were "joined in matrimony by Sacramento Municipal Court Judge Peter Manino.

Some 700 Expo visitors watched from the grandstand. The bride, of Inglewood, was attired in an abbreviated white deerskin costume and carried a bouquet of white carnations. The bridegroom, of 7521 Woodley Ave., Van Nuys, wore his traditional Indian garb and carried a tomahawk."

According to the article "A dozen Hollywood stuntmen, all in western dress, participated in the ceremony, which was conducted on a stage amid special props used by the stuntmen in their show."

A photographer who attempted to record the event stepped onto a trick breakaway table and it collapsed under him. "He took the fall like a veteran stunt performer, but his pants didn't."

The article states that Becky "arrived in a century-old King George IV marriage phaeton borrowed from a Cal Expo Carriage Rally and walked to the altar under a bridge of Bowie knives extended by the stuntmen."

Judge Mannino asked the traditional question if there were any objections to the wedding.

A stuntman appeared atop a saloon set, shouting that he had some. A shot rang out. He gasped and di a midair somersault onto a mattress 20 feet below.

Two other stuntmen protested from roofs of other Frontier St. buildings and were immediately toppled from their perches onto the street below. "If all objections have been removed…" the judge said, and the ceremony continued.

The bride, who attended Hawthorne High School as Becky Downing, and the groom, a graduate of Alhambra High School whose legal name is Skeeter Vaughan, exchanged wedding bands and rode off together on a white horse, as the audience threw rice in their path. Vaughan is half-Cherokee. Blue-eyed Becky is one-quarter Cherokee."

The married couple returned moments later for the reception. "The bride held the wedding cake – shaped like a tepee – and the groom cut it, at 10 paces, with a tomahawk.

Then, with their tomahawk and Bowie knife throwing act, coming up within the hour, Grey Otter and Becky Little Eagle took a quick honeymoon cruise down Cal Expo's lagoons in a paddleboat."

In the Daily Times News, February 4th 1970 an article discusses Rebecca Little Eagle Grey Otter. It states that she "is probably the only wife in the world who couldn't get a divorce by testifying that her husband throws knives and hatchets at her.

In Lawrence Truman's screen version of the Broadway hit "The Great White Hope," Beckie stands calmly against a wall while a Cherokee brave, who happens to be her husband, rings her head with tomahawks tossed from 30 feet. For this, she and her mate, Skeeter Vaughan Grey Otter, receive together something like $1000 a week.

The job on the 20th Century-Fox movie, which headlines James Earl Jones in the role of the first black heavyweight boxing champ, ran just a week.

"But we're always busy," says Skeeter. "We work county and state fairs and special events all the time."'

The article states: "the Grey Otters are both stunt specialists in motion pictures and television. Aside from his prowess with the tomahawk. Skeeter is in constant demand for falls, both from galloping horses and high cliffs.

His most difficult feat, he says, was riding a horse and firing a shotgun blast so powerful it supposedly hurled him out of the saddle.

"I had a spring harness hooked onto my back under my costume. When my shotgun went off someone pulled the lever which activated the spring and yanked me off the horse backwards. I flew through the air and landed on my rear. It was hard ground. That one hurt."

Skeeter was paid $200 for 10 minutes work—but nobody disputed that he earned every nickel. Generally, stunt fees are negotiable and depend on the skill and risk involved.

Nevertheless, he prefers his tomahawk work above anything else in his stunting book. In 'The Great White Hope," his chore was simple – for him – and he didn't even practice to get in shape. Director Martin Ritt required several takes of the scene, which shows the carnival atmosphere attending the big championship bout in Reno, with Jones (impersonating Jack Johnson) entering the stadium. As the cameras turned and the extras roared, Skeeter calmly and repeatedly crowned his bride with those hatchets."

The article states that "In other films, however, he has tossed sharp weapons at such luminaries as Randy Boon and James Drury ('The Virginian'), Peter Breck ("The Big Valley"), Connie Hines ("Mr. Ed") and among others. None used a stunt double, Skeeter proudly relates. They all trusted his marksmanship. At 50 feet, he has the accuracy of a pistol shot, he claims.

Mrs. Grey Otter was asked if she flinches or closes her eyes when the missiles are thrown.

"'Oh, No,' she replied. "I love to watch the audience and note their reaction. The worst part of it is the sound. When the ax hits the board within two or three inches of my head it makes a terrible loud noise."

The Grey Otters have only one rule. He never throws at her after they've had a family argument. "This forces us to have a harmonious life together," he smiled at her. "We can't afford to fight."

THE GREEN TRIANGLE
(Through the Courtesy of Bob Hillman)

Presents...
GREY OTTER

WORLD'S GREATEST PRECISION KNIFE AND TOMAHAWK THROWER

Direct from Universal and C.B.S. Studios of Hollywood, California.

Grey Otther will perform the World's Most Dangerous Act for your enjoyment at The Green Triangle on SATURDAY and SUNDAY, OCTOBER 7TH & 8TH.

Grey Otter has appeared in such films as Geronimo, Melody Ranch, Gunsmoke, Daniel Boone, Laredo, The Man From U.N.C.L.E., and many others, and has appeared with such film notables as Jack Palance, Connie Hines, Randy Boone, James Drury, Peter Breck, Paula Lane, Robert Easton, Dale Robertson, Roy Rogers, and Dale Evans. Grey Otter will be assisted in his performance by the lovely Miss Megan Morgan. Grey Otter is one of the many variety artists made available through the Universal Talent Agency of Beverly Hills, California, and Pocatello, Idaho, which is managed in this area by Richard L. Armstrong and Robert W. Bevans.

Make Reservations Early
For Your Christmas Parties
Reserve Now and Be Assured of the
Date You Prefer

Grey Otter's passing in Sequim, Washington, was announced in the San Bernardino County newspaper, Monday March 13th 1989, he had appeared in more than 100 films, television shows and commercials. He died of a heart attack at the age of 66.

He died "at Olympic Memorial Hospital in Port Angeles, according to the Seattle talent agency that last represented him. He had lived in Sequim since August. His film credits included "The Last of the Mohicans," "Revolution," and "Like Father Like Son."

When I interviewed members of his family it left me with the impression that Grey Otter did what he could to make a name for himself, and he provided for his family doing what he did best, throwing knives and tomahawks and doing working as a stuntman.

He was a proud Veteran, he survived a war, he also helped others to learn some skills in order that they too might survive a war. But unfortunately later in life some people took advantage of him.

There are people in this world who take advantage of others. It's one of the sad facts of life. Sometimes the people who are taken advantage of were the people that protected us in life, they fought for our Country so that we would remain free. Grey Otter had associates that didn't have his best interests at heart. I was informed that after Grey Otter died his house was ransacked by these associates.

I don't know who those associates were, but in correspondence with his family I got the impression that these were people who were involved with him in show business or his knife and tomahawk throwing endeavors. They were probably business associates. It was upsetting for his family that this occurred as it would be for anybody who has ever had something like that happen to a loved one. I don't know the full extent of it but for the purposes of this book that is probably all that needs to be said regarding that subject, it is unfortunate that occurred.

What It Means To Me

I am not a Native American, nor am I a soldier or someone who might need to know a tactical tomahawk fighting system, or any unarmed combat or martial arts system for that matter, in order to survive some type of conflict. I learn hand to hand combat systems not because I ever expect to use them. I hope I am never in a position where I might have no option but to use them. The truth is I no longer learn these systems for self-defense purposes. I mainly learn the techniques for posterity, and because of my interest in history.

Some of the other answers as to why I have been interested in exploring and understanding, the history and techniques of "Martial Arts" and reality based systems are found in an article written by George Leonard, for Esquire magazine in July 1986.

Discussing action heroes of that time period such as Rambo, Leonard writes "America has discovered a new hero, the latest in a lineage that goes back to Davy Crockett and Daniel Boone, to the Lone Ranger and the western marshal with the fast draw. This new hero, like his predecessors, is always on the side of Right, but not necessarily on the side of the Establishment. Unlike the World War II team player, he is a lone fighter, a common man who through strenuous self-discipline and rigorous training has developed extraordinary skills, which he puts to use with devastating results."

Unlike Leonard I was a kid when all these action films were being made and released. They did have a profound impact on me. Like other kids my age I always found myself rooting for the action hero who was typically fighting Commies or Terrorists and kicking major ass. The thing is, that was the Hollywood portrayal, the reality was when I met people from the military or law enforcement they weren't anything like this. In my experience Veterans were typically very humble, very down to earth people.

Leonard begs the question, what if there was no Rambo? Who then? That question is still very relevant today. He writes "It is a question that tends to paralyze our mental processes. For many of us who are dedicated to peace, the very idea of the "good" warrior seems a contradiction. We are haunted by images of armed soldiers in a city square, of innocent people kidnapped, tortured, or made to "disappear." The word "military" can conjure up the word "dictatorship." The word "police" joins all too easily with "state."

Leonard continues "Still, in this violent dangerous world, only the most fevered idealist would dispense with soldiers and policemen. So the question remains… If we are going to have people whom we give the job of risking their own lives and, if necessary, taking the lives of others, how are we to deal with them? How are we to think about them? And beyond that, is there some way that the warrior spirit at its best and highest can contribute to a lasting peace and to the quality of our individual lives during the time of peace?" It's an interesting question given today's world, a lot has changed since the article was written in the 1980s.

Leonard writes "I approach these questions not as a distant, dispassionate observer, but as one who served as a combat pilot in the south-west pacific in World War II and as an air-intelligence officer during the Korean War. More recently, I've spent fifteen years studying a martial art called aikido, one dedicated to harmony, but a martial art nonetheless, with roots that go back to the medieval Japanese samurai. Through my association with this art, I've developed training programs and simulation games designed to produce the warrior spirit in men and women who never plan to go to war."

The interesting thing in the article is that Leonard quotes Don Casteneda, someone Swiftdeer had claimed was one of his teachers. Leonard writes "In 1963, Castaneda, an anthropology student, became the apprentice of a Yaqui Indian shaman named Don Juan Matus, who lived in the northern Mexican desert. His books, which include The Teachings of Don Juan, A Separate Reality, and Tales of Power, describe the adventures and ordeals of his apprenticeship.

To become a "man of knowledge," Don Juan tells Castaneda, it is necessary to be a warrior. A warrior is not one who goes to war and kills people, but rather one who exhibits integrity in his actions and control over his life. The warrior's courage is unassailable, but even more important are his will and patience. He lives every moment in full awareness of his own death, and, in light of this awareness, all complaints, regrets, and moods of sadness or melancholy are seen as foolish indulgences.

Don Juan's warrior pursues power and acts strategically in order to achieve self-mastery. "The spirt of a warrior is not geared to indulging and complaining, nor is it geared to winning or losing. The spirit of a warrior is geared only to struggle.... Thus the outcome matters very little to him." The warrior aims to follow his heart, to choose consciously the items that make up his world, to be exquisitely aware of everything around him, to attain total control, then act with total abandon. He seeks, in short, to live an impeccable life."

According to Leonard "Castaneda's notion of the warrior resonates with ancient echoes. *Almost every culture has had its own version of an ideal warrior's code. It exists in its purist form, unwritten, among peoples we call primitive – American Indians, African tribesmen.* It has often been honored in the breach, especially in the nation-states of the epoch we call civilization. Still, it remains an ideal to be realized, a guide to living that might prove useful in today's complex and vexing world."

Leonard goes on to discuss meeting a man named Donald Levine, a professor of sociology and dean of the College at the University of Chicago. "He is also a dedicated martial artist. I had wanted to meet him ever since reading a short version of his article "The Liberal Arts and the Martial Arts."

Leonard liked the article because he thought it "went a long way toward clarifying the role of the warrior in a free society. In it, he defines the liberal arts as including all education that is undertaken for self-development, all learning that exists essentially for its own sake rather than for some utilitarian purpose. Liberal education, according to Levine, first emerged in two unique cultures, those of classical Greece and China. IN both of those cultures, such education was considered the highest human activity. And, though it might seem strange in light of today's academic climate, it included the cultivation of combat skills as well as intellectual skills. In both the East and West, in other words, the martial arts and the liberal arts arose together, and were equally revered.

In the centuries that followed, this ideal was often lost. Both the art of combat and the education of the intellect were at times corrupted and put to narrow and exploitative uses. But during certain creative moments in history – for example, when the Buddhist monk Bodhidharma introduced Ch'an (Zen) Buddhism and the forerunner of Shaolin temple boxing to China in the sixth century A.D. – the liberal education of both body and mind has flourished."

After reading his article Leonard met and trained with Levine and afterwards they went to dinner and discussed his ideas. He asked Levine "What about now? Do the martial arts have anything significant to offer late-twentieth-century America?

Levine's answer was "Yes, I can see this as a time when the body and mind are being reunified, a time when the liberal arts can learn a great deal from the martial arts. This is true, of course, only when the martial arts are practiced primarily for mastery of their intrinsically beautiful forms and for self-development rather than primarily for self-defense or for the brutal sensation you see in the movies. And arts like aikido, which tie ethical vision right into daily practice, are just what this country needs. Remember what the founder said: the point of aikido training is to create persons who evince 'a spirit of loving protection for all beings, who bind the world together in peace and unity."

I understand what Levine is saying to a certain degree. I am learning boxing these days. It is one of the world's first "martial" arts. I go to the gym and train for self-development. Initially I was interested in learning it for self-defense but now I am there learning it not because I want to be a champ or anything but because I want to develop myself. I also still train in reality based self-defense systems as well but that is more about training the "self-defense" aspects. I take a covers all bases approach with my training. One thing about the article is that Brazilian Jiu Jitsu and Mixed Martial Arts have overtaken other traditional martial arts systems such as Aikido, and that is typically what potential students look for today.

Leonard writes "The heart of this way of life is practice itself, the regular, systematic, unremitting practice of the dedicated martial artist. And then there is a progression of learning common to the martial arts that leads to the transcendence of mere technique. 'One begins by self-consciously practicing a certain technique.'"

Levine had written in his article "One proceeds slowly, deliberately, reflectively; but one keeps on practicing, until the technique becomes internalized and one is no longer self-conscious when executing it. After a set of techniques has been thoroughly internalized, one begins to grasp the principles behind them. And finally, when one has understood and internalized the basic principles, one no longer responds mechanically to a given attack, but begins to use the art creatively and in a manner whereby one's individual style and insights can find expression." Leonard writes "A fine way of learning for the scholar – and for the warrior." I find this assessment to be true of both reality based self-defense systems as well as sportive systems. One needs to continuously practice them in order to master the techniques so that they become second nature.

Leonard then discusses how he made the jump from Rambo to the Dean of a prestigious college in search of the warrior ideal. He writes that he had "the opportunity to spend two days with twenty-four real-life Rambos, and discovered that the stretch was not as great as might be imagined."

Leonard goes on to write: They were Green Berets, members of the U.S. Army Special Forces, who had volunteered for an experimental six-month course in advanced mind-body training run by a Seattle-based organization called SportsMind. Most of them had gone through Army Ranger training. All were skilled in hand-to-hand combat and the use of various weapons, parachuting scuba diving, rock climbing, skiing, escape and evasion, and other specialized military skills, some of them classified. The experimental training program, designed to add a psychological component to an already rigorous schedule of military training, including daily aikido training aimed at integrating the physical and the mental

Jack Cirie, who had recently retired from the Marines as a lieutenant colonel, led the training. Richard Heckler, who is a Ph.D. in psychology as well as a gifted aikidoist, was engaged for both his psychology and martial-arts skills. I was one of several consultants called in during the six months of the program; my job was to serve as a guest aikido instructor, and to lecture on challenge and change as expressed in two of my books, which were on the trainee's reading list.

I met the Special Forces men at a small, unused base theater that had bene converted to a dojo. They were dressed in martial-arts uniforms; the only concession to military dress was the presence of olive-drab name patches sewn to the uniforms above the left breast. The men ranged in age from twenty-two to forty-one, and in rank from buck sergeant to captain. But age and rank held little significance as they kneeled at the edge of the mat. As is customary in aikido training, I knelt in front of them, facing a photograph of the master who founded the art. I bowed, straightened up, and then clapped twice sharply, a traditional gesture signaling readiness. My new students clapped along with me, and I could sense their power and decisiveness.

I started with warm-up exercises, followed by some gentle stretching, then demonstrated the first martial-arts technique. As the men paired off and took turns attacking each other, I moved from one to the other, making suggestions, providing individual demonstrations.

It was quickly apparent that these elite troopers were expert learners. This should have come as no surprise. The peacetime military is primarily a gigantic educational institution, and most military men today spend most of their time learning new skills and honing those they already know. I could spot a certain amount of kidding around, and anything that wasn't fully understood was quickly challenged. But these were students any teacher would love to teach. They were fiercely attentive. They worked hard. They were willing to try anything. They were exceptionally eager to master each technique.

At the same time, these soldiers exhibited a sense of courtesy and respect in their relationship with me that seemed neither forced nor pro forma. And, though I knew they were superb fighting men, I saw in them none of the gratuitous brutality that marks the cinematic version of the Special Forces trooper. At one point during my lecture, I asked how many of them felt that Rambo accurately represented the Special Forces soldier. Only one man - the group jokester- raised his hand. Then I asked how many had enjoyed the movie. Most raised their hands. They had liked the action. But one man told Jack Cirie that unless Rambo started carrying his "ruck,' he wasn't going to see any more of his movies. The ruck, or rucksack, is the symbolic and literal mark of the real Green Beret. Unless you've paid your dues by humping a hundred or so kilometers with eighty or ninety pounds in your ruck, you're just a Hollywood warrior.

How did they define the ideal warrior? It was a subject I kept bringing up during informal talks, a subject that also had been discussed in previous classroom sessions. It appeared that these men's definition was not far removed from that of the perennial warrior's code. They cited loyalty, patience, intensity, calmness, compassion, and will. They agreed that the true warrior knows himself, knows his limitations. "It's not that you don't have holes," Cirie said, "It's that you're aware of the holes."

Self-mastery, according to the Special Forces men, is a warrior's central motivation. He is always practicing, always seeking to hone his skills, so as to become the best possible instrument for accomplishing his mission. The warrior takes calculated risks and tests himself repeatedly. He works well within a group but also is a self-starter. He believes in something greater than himself: a religion, a cause. He does not worship violence but is at home with it. He is human, not a robot. He may snivel (their word for complain), but he is not a victim. One top sergeant, who had been in Vietnam, said, "We're all acolyte warriors until we've been tested in combat." But others felt that the warrior could exist even outside of the military.

What most struck me was the importance these elite soldiers placed on service and protection. Again and again this subject came up in our conversations, not only as a warrior ideal, but also as a compelling justification for their way of life itself. "These guys," Heckler said to me in a crowded restaurant, "genuinely feel they're protecting everybody in this room."

It's important to note that this article was written in the 1980s and a lot has changed since then. The author of the article and some of the people mentioned in it have also passed away. In some ways some things have stayed the same.

Leonard writes "It would be just as much a mistake to glorify and to denigrate the servicemen, or the warrior ideal. But in a culture where million-dollar lottery winners are accorded headline glory, where putting together an essentially destructive stock deal is considered a heroic act, where the Good Life is tied in with getting a corner office and driving a certain make of German car, and where "the one who dies with the most toys wins" is offered as gospel truth rather than as a sick joke, I found it refreshing to have met people who hold alternative views and live by different precepts."

He writes: "We've learned that the military and police forces possess great power to oppress as well as protect, and it's clear that for a free society to survive, they must be thoroughly depoliticized and placed firmly under civilian authority and review. It's also clear that armed forces can become so overarmed and eager for action that they can provoke conflict rather than promote peace. To beat swords into plowshares, especially in a nuclear age, remains one of the highest human endeavors. Meanwhile, it seems obvious that as long as wars of any sort exist, it's better to have good soldiers than bad ones. Rambo won't do. He's too sullen, headstrong, self-centered, delinquent, and -face it- unreal; he doesn't carry his ruck. Nor do we need the generally two-dimensional, brutish, bullet-spraying locoes of other action films, or the real-life fantasists who buy exotic weapons and try to cover their inadequacies with camouflage cloth. We need military men and women who are effective, who are professional, who live by a spoken or unspoken warrior's code, and who are dedicated to keeping the peace. Such men and women do exist. They don't deserve to be represented as distorted superheroes. They do deserve to be acknowledged and appreciated for what they are.

And what about war itself? In his seminal book The Warriors, philosopher J. Glenn Gray, a World War II combat veteran, writes, "No human power could atone for the injustice, suffering, and degradation of spirit of a single day of warfare." At the same time, he reminds us of war's terrible and enduring appeal: the opportunity to yield to destructive impulses, to sacrifice for others, to live vividly in the moment. The appeal of war is not a popular subject, but until we deal with it openly and undogmatically we may never find a warrior's path toward peace.

One friend of mine – a peace-loving man who served as a medic with General George S. Patton's forces as they fought their way across Germany – has told me the early-morning smell of a cup of coffee in a snow-covered German forest is more real, even now, forty years later, than anything in his present surroundings. And the unbelieving, strangely amused look on the pilot's face in the plane next to me as his windscreen was shattered by ground fire just north of Manila remains as marvelously crystal-clear today as it was then. "We do not know," Gray writes, "whether a peaceful society can be made attractive enough to wan men away from the appeals of battle. Today we are seeking to make war so horrible that men will be frightened away from it. But this is hardly likely to be more fruitful in the future than it has been in the past. More productive will certainly be our efforts to eliminate the social, economic, and political injustices that are always the immediate occasion of hostilities. Even then, we shall be confronted with the spiritual emptiness and inner hunger that impel many men toward combat. Our society has not begun to wrestle with this problem of how to provide fulfillment to human life, to which war is so often an illusory path."

Leonard writes: "I've come to believe that Gray is right. The problem is not that war is so often vivid, but that peace is so often drab. Looking at this same problem back in 1910, the psychologist William James argued that we need "a moral equivalent of war," a way of living that would provide the challenges of combat without its horrors. James's argument, it seems to me, becomes more compelling with the development of each new weapons system. In this light, peace advocates are indeed doing important work in opposing war through public statements, petitions, and demonstrations. But the end of war – can we imagine it? – might require something more fundamental: the creation of a peace that is not only just, but also vivid.

The work of creating a more vivid peace must address the problem of our spiritual emptiness and inner hunger. It might well require that we relinquish some of our currently fashionable cynicism and give more energy, as Gray suggests, to values that could be called moral and spiritual. But there's something else: We need passion. We need challenge and risk. We need to be pushed to our limits. And I believe this is just what happens when we accept a warrior's code, when we try to live each moment as a warrior, whether in education, job, marriage, child rearing, or recreation. The truth is that we don't have to go to combat to go to war. Life is fired at us like a bullet, and there is no escaping it short of death. All escape attempts – drugs, aimless travel, the distractions of the media, empty material pursuits – are sure to fail in the long run, as more and more of us are beginning to learn.

Could it be that the current popularity of Rambo and the other warrior films goes beyond neopatriotism and revenge fantasies? Let's at least consider the possibility that the warrior rage also signifies something we haven't yet put into words: that there are many potential warriors among us, that each of us, at some level, wants to meet life head-on, to risk everything for what he believes in, to develop himself to the fullest, and to serve others.

When the samurai Kikushi was ordained a bodhisattva (one devoted to lifelong service), his master told him "You must concentrate upon and consecrate yourself wholly to each day, as though a fire were raging in your hair." On those frequent occasions when statements like this sound hopelessly overblown and quite impossible to achieve in real life, I recall something Jack Cirie said during one of our conversations: "Believing you can be perfect is the fatal imperfection. Believing you're invulnerable is the ultimate vulnerability. Being a warrior doesn't mean winning or even succeeding. But it does mean putting your life on the line. It means risking and failing and risking again, as long as you live."

My hope is that by writing this book I provided Grey Otter and other warriors like him some exposure by reintroducing them and some of their accomplishments. I hope that I was able to bring some necessary exposure to some of the other great Native American veterans from WW2 such as Miles Horn who also did his part to help others and who shared his knowledge of the warrior arts in order that others might survive.

It was important to discuss Brummett Echohawk who like Miles Horn was a celebrated artist and did his best to expose certain aspects of Native American "warrior" life in order to help others have a better understanding of all that entailed.

It was important to mention Swiftdeer who attempted to bring some exposure to the Native American warrior culture, whether it was ultimately accepted or not. He did his part in the best way he knew how, through teaching the martial arts.

I could have mentioned the Navajo Code Talkers and others but they have received some well-deserved exposure already. It was important to mention the ones who may not have had as much exposure. These people also mattered and they made a difference. It is my hope that I was able to convey that with this book.

About The Author

Mr. Sabet has been interested in the Martial Arts since he was a young boy growing up as an expat in Thailand. He was always interested in History and loved to hear all the stories that the old-timers had to tell. As a teenager he got into Rock music and started one of the first Punk bands in Thailand with his friends.

When he went off to College in Washington State, he stuck to his interests, taking lessons in Judo and playing in loud bands. After College and having worked a few dead end jobs, he decided it was time to move to NYC to be closer to family, where he experienced all that the city had to offer, including the horrific 9/11 attacks. As the years went by he continued to dabble in different martial arts and to research various stories, most of which were local interest related.

One such story that has taken him several years to research and write, is a True Crime story about a bank robbery which occurred in the 1950s. Through the course of his study, he discovered that some of the criminals involved had been Veterans of the 2nd World War. This discovery led him to delve deeper into what particular skill-set a veteran might have and utilize for good or bad after returning from a War.

It was during this process of discovery that Mr. Sabet had arrived at the knowledge of WWII Era Hand to Hand Combat.

Made in the USA
Las Vegas, NV
23 March 2023